The Handbook of Environmental Chemistry

Volume 3   Part E

Edited by O. Hutzinger

Anthropogenic Compounds

With contributions by
F. Adams, S. J. Blunden, R. van Cleuvenbergen, C. J. Evans,
L. Fishbein, U.-J. Rickenbacker, Ch. Schlatter, A. Steinegger

With 23 Figures and 42 Tables

Springer-Verlag
Berlin Heidelberg GmbH

Professor Dr. Otto Hutzinger
University of Bayreuth
Chair of Ecological Chemistry and Geochemistry
Postfach 10 12 51, D-8580 Bayreuth
Federal Republic of Germany

ISBN 978-3-662-15069-6

Library of Congress Cataloging in Publication Data
(Revised for vol. E)
Anthropogenic compounds.
(The Handbook of environmental chemistry; v. 3) Includes bibliographical references and indexes. 1. Pollution —
Environmental aspects. 2. Pollutants — Toxicology. 3. Environmental chemistry. I. Anliker, R. (Rudolf), 1926–
II. Series.
QD31.H335 vol. 3   574.5′222 s [574.5′222]   80–16609 [QH545.Al]
ISBN 978-3-662-15069-6   ISBN 978-3-540-46211-8 (eBook)
DOI 10.1007/978-3-540-46211-8

© Springer-Verlag Berlin Heidelberg 1990
Originally published by Springer-Verlag Berlin Heidelberg New York in 1990
Softcover reprint of the hardcover 1st edition 1990

2152/3020-543210

Preface

Environmental Chemistry is a relatively young science. Interest in this subject, however, is growing very rapidly and, although no agreement has been reached as yet about the exact content and limits of this interdisciplinary subject, there appears to be increasing interest in seeing environmental topics which are based on chemistry embodied in this subject. One of the first objectives of Environmental Chemistry must be study of the environment and of natural chemical processes which occur in the environment. A major purpose of this series on Environmental Chemistry, therefore, is to present a reasonably uniform view of various aspects of the chemistry of the environment and chemical reactions occurring in the environment.

The industrial activities of man have given a new dimension to Environmental Chemistry. We have now synthesized and described over five million chemical compounds and chemical industry produces about one hundred and fifty million tons of synthetic chemicals annually. We ship billions of tons of oil per year and through mining operations and other geophysical modifications, large quantities of inorganic and organic materials are released from their natural deposits. Cities and metropolitan areas of up to 15 million inhabitants produce large quantities of waste in relatively small and confined ares. Much of the chemical products and waste products of modern society are relased into the environment either during production, storage, transport, use or ultimate disposal. These released materials participate in natural cycles and reactions and frequently lead to interference and disturbance of natural systems.

Environmental Chemistry is concerned with *reactions in the environment*. It is about distribution and equilibria between environmental compartments. It is about reactions, pathways, thermodynamics and kinetics. An important purpose of this Handbook is to aid understanding of the basic distribution and chemical reaction processes which occur in the environment.

Laws regulating toxic substances in various countries are designed to assess and control risk of chemicals to man and his environment. Science can contribute in two areas to this assessment; firstly in the area of toxicology and secondly in two areas of this assessment: firstly in the area of toxicology and secondly in the area of chemical exposure. The available concentration ("environmental exposure concentration") depends on the fate of chemical compounds in the environment and thus their distribution and reaction behaviour in the environment. One very important contribution of Environmental Chemistry to the above mentioned toxic substances laws is to develop laboratory test methods, or mathematical correlations and models that predict the environmental fate of new chemical compounds. The third purpose of this Handbook is to help in the basic understanding and development of such test methods and models.

The last explicit purpose of the handbook is to present, in a concise form, the most important properties relating to environmental chemistry and hazard assessment for the most important series of chemical compounds.

At the moment three volumes of the Handbook are planned. Volume 1 deals with the natural environment and the biogeochemical cycles therein, including some background information such as energetics and ecology. Volume 2 is concerned with reactions and processes in the environment and deals with physical factors such as transport and adsorption, and chemical, photochemical and biochemical reactions in the environment, as well as some aspects of pharmacokinetics and metabolism within organisms. Volume 3 deals with anthropogenic compounds, their chemical backgrounds, production methods and information about their use, their environmental behaviour, analytical methodology and some important aspects of their toxic effects. The material for volumes 1, 2, and 3 was more than could easily be fitted into a single volume, and for this reason, as well as for the purpose of rapid publication of available manuscripts, all three volumes are published as a volume series (e.g. Vol. 1; A, B, C). Publisher and editor hope to keep the material of the volumes 1 to 3 up to date and to extend coverage in the subject areas by publishing further parts in the future. Readers are encouraged to offer suggestions and advice as to future editions of "The Handbook of Experimental Chemistry".

Most chapters in the Handbook are written to a fairly advanced level and should be of interest to the graduate student and practising scientist. I also hope that the subject matter treated will be of interest to people outside chemistry and to scientists in industry as well as government and regulatory bodies. It would be very satisfying for me to see the books used as a basis for developing graduate courses on Environmental Chemistry.

Due to the breadth of the subject matter, it was not easy to edit this Handbook. Specialists had to be found in quite different areas of science who were willing to contribute a chapter within the prescribed schedule. It is with great satisfaction that I thank all authors for their understanding and for devoting their time to this effort. Special thanks are due to the Springer publishing house and finally I would like to thank my family, students and colleagues for being so patient with me during several critical phases of preparation for the Handbook, and also to some colleagues and the secretaries for their technical help.

I consider it a privilege to see my chosen subject grow. My interest in Environmental Chemistry dates back to my early college days in Vienna. I received significant impulses during my postdoctoral period at the University of California and my interest slowly developed during my time with the National Research Council of Canada, before I was able to devote my full time to Environmental Chemistry in Amsterdam. I hope this Handbook will help deepen the interest of other scientists in this subject.

Otto Hutzinger

This preface was written in 1980. Since then publisher and editor have agreed to expand the Handbook by two new open-ended volume series: Air Pollution and Water Pollution. These broad topics could not be fitted easily into the headings of the first three volumes.

All five volume series will be integrated through the choice of topics covered and by a system of cross referencing.

The outline of the Handbook is thus as follows:
1. The Natural Environment and the Biogeochemical Cycles
2. Reactions and Processes
3. Anthropogenic Compounds
4. Air Pollution
5. Water Pollution

Bayreuth, September 1989   Otto Hutzinger

Contents

List of Contributors

Prof. Freddy Adams
Department of Chemistry
University of Antwerp
Universiteitsplein 1
B-2610 Wilrijk

Dr. Stephen J. Blunden
International Tin Research Institute
Kingston Lane, Uxbridge
Middlesex UB8 3PJ
England

Prof. Rudy van Cleuvenbergen
Department of Chemistry
University of Antwerp
Universiteitsplein 1
B-2610 Wilrijk

Dr. C. J. Evans
International Tin Research Institute
Kingston Lane, Uxbridge
Middlesex UB8 3PJ
England

Dr. Lawrence Fishbein
ILSI
1126 Sixteenth Str.
Washington, DC 20036
UK

Dr. Urs-Josef Rickenbacher
Sandoz AG
P.O. Box
CH-4002 Basel

Dr. Christian Schlatter
Institut of Toxicology,
ETH and University Zurich
Schorenstr. 16
CH-8603 Schwerzenbach

Dr. Alfred Steinegger
Schweizerische Aluminium AG
Feldeggstr. 4
CH-8034 Zürich

Organotin Compounds

Stephen J. Blunden and Colin J. Evans

International Tin Research Institute, Kingston Lane, Uxbridge, Middlesex UB8 3PJ.

Summary

The impact of organotin compounds in the environment is assessed, based on a critical survey of the published literature. After a brief introduction, the properties and characteristics of organotin compounds are outlined, with notes on methods of manufacture, patterns of use and chemical properties. A section deals with analytical methods used to determine low levels of organotins in aqueous systems, in soils and sediments, in air samples and in biological materials, The next section examines sources of entry of organotin compounds into the environment via the principal areas of use, namely, PVC stabilisation, wood preservation, agricultural pesticides and marine antifouling coatings; possible effects of the introduction of organotin molluscicides are examined briefly. An important section is concerned with environmental interactions which are linked closely with the possible long-term hazards posed by these compounds. Relevant to this section are the aqueous chemistry of organotins, environmental methylation and mechanisms of degradation (particularly by UV irradiation and biological cleavage). The toxicology of organotin compounds is considered in some detail and results are summarised for animal studies and for reported effects in humans; the mode of toxicological action, metabolic pathways and results of carcinogenicity testing are also described in this section. Preventive measures are described in the next section and include methods of waste disposal. The final section look sat future trends, bearing in mind the growing industrial use of organotin compounds (currently 35,000 tonnes/year).

Introduction

Credit for the first systematic study of organotin compounds belongs to Sir Edward Frankland (1825–1899), who in 1853 prepared "di-iodo diethylstannane" (diethyltin di-iodide) and in 1859 "tetraethylstannane" (tetra-ethyltin); other compounds followed. These organotins remained no more than laboratory curiosities for nearly 100 years since they did not reveal any obvious commercial potential. Their chance came in the 1940s when the plastics industry began to expand and the value of polyvinyl chloride (PVC) began to be realised. This polymer suffered from the disadvantages that, under the effects of heat or light, it degraded with consequent embrittlement and discolouration. A search was made for compounds which would inhibit this degradation and the powerful stabilising action of certain organotin derivatives was disclosed. They were introduced for this purpose in the USA in the late 1940s and in Europe in the mid-1950s. A breakthrough in their use occurred in 1955 when Government approval was obtained in the USA and elsewhere for the use of non-toxic octyltin stabilisers in PVC intended for food-contact applications, for example in PVC bottles for fruit squashes or vegetable oils. The stabilisation of PVC has remained the largest single application of organotin compounds up until the present day.

Biocidal uses of organotin derivatives stemmed from a systematic study which was sponsored in the 1950s by the International Tin Research Council at the Institute for Organic Chemistry, TNO, Utrecht [143, 144, 226–231]. The powerful biocidal properties of the trialkyltin and triaryltin compounds were discovered and it was realised that these characteristics could be exploited commercially. Triphenyltin compounds were found to have fungicidal activity combined with very low phytotoxicity and in the late 1950s triphenyltin acetate was introduced commercially as an agrochemical, proving particularly effective in protecting potatoes against *Phytophthora infestans*. Shortly afterwards, a product based on triphenyltin hydroxide was marketed for similar applications. Since the late 1960s, three other organotin compounds have been introduced for use as miticides, protecting orchard fruits and a

Fig. 1. Organotin compounds are used in a very wide range of commercial products.

host of other crops against the ravages of these pests. Other biocidal uses of triorganotin compounds which have become widespread include wood preservation and antifouling paints; smaller outlets are disinfectants and algicidal treatments for building materials. Current consumption of organotin compounds is believed to be about 35,000 tons/year and this may increase considerably since new applications seem likely as a result of research currently in progress. The development of the organotin chemical industry has been reviewed in more detail by Bennett [35].

It can be seen that a wide variety of properties is exhibited by organotin compounds and this is reflected in the varied industries in which they are used (Fig. 1). Environmental control thus necessitates an understanding of the technologies involved in each application and an assessment of the importance of organotin compounds in each area. In some cases the introduction of these compounds has had the effect of reducing long-term hazards, where the fact that organotin pesticides eventually break down to harmless non-toxic forms of tin, is an advantage. In the case of PVC stabilisation, the use of non-toxic organotin compounds have enabled PVC bottles to retain their glass-clear clarity on the supermarket shelf.

Properties and Characteristics of Organotin Compounds

Organotin compounds are characterised by the presence of one or more tin-carbon bonds and have the general formula R_nSnX_{4-n}, where R is an alkyl or aryl group, X is an anionic species, eg. chloride, oxide, hydroxide or other functional group, and n is 1 to 4.

Manufacture

Methods of manufacture of organotin compounds usually comprise two principal steps; the first consists of making direct tin-carbon bonds in compounds such as R_4Sn; the second stage is one of coproportionation in which R_4Sn is reacted with stannic chloride to produce compounds of the type R_3SnCl, R_2SnCl_2 and $RSnCl_3$. Other derivatives may then be simply produced from these chlorides, for industrial end uses. Reaction schemes are shown in Fig. 2. Detailed accounts of the chemistry [89] and the process technology [49] have been published elsewhere.

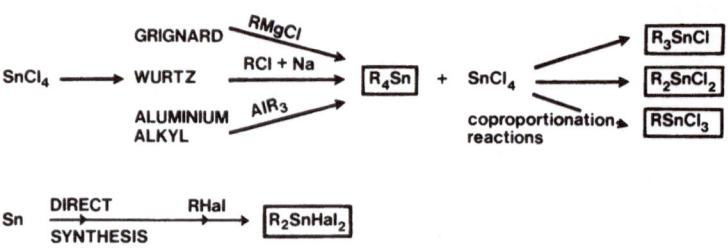

Fig. 2. Manufacturing routes to organotin compounds.

The Grignard route for the industrial production of tetraorganotins is flexible and produces high yields; it is used to manufacture tetraphenyltin, tetrapropyltin, tetra-butyltin and tetraoctyltin compounds. In the Wurtz synthesis sodium is used in place of magnesium; competing side reactions are a problem and to date this process is only used commercially in East Germany. The aluminium alkyl route has advantages in that it can be operated continuously and no solvents are needed. It is used in FR Germany to produce tetraalkyltins, particularly tetraoctyltin. There is also a direct synthesis route which involves reaction between tin and alkyl halides. Iodides are needed to achieve a suitably high reaction rate so that the process is only industrially viable when the iodine can be recovered.

In the coproportionation stage, redistribution between the tetraorganotin and tin (IV) chloride occurs:

$$3R_4Sn + SnCl_4 \rightarrow 4R_3SnCl$$
$$R_4Sn + SnCl_4 \rightarrow 2R_2SnCl_2$$
$$R_4Sn + 3SnCl_4 \rightarrow 4RSnCl_3$$

Patterns of Use

The number of Sn—C bonds has a profound effect on the properties of organotins allowing a range of applications (Table 1). Tetraorganotins R_4Sn, are usually colour-

Table 1. Principal industrial uses of organotin compounds

Application	Function	Principal compounds used
PVC stabilisation	Stabilisation against effects of heat and light	dialkyltin di-isooctylthioglycolate (alkyl = methyl, butyl, octyl, 2-butoxy-carbonylethyl) dialkyltin maleate (alkyl = methyl, butyl, octyl) mono-alkyltin tri-isooctylthio-glycolate (alkyl = methyl, butyl, octyl, 2-butoxy-carbonylethyl)
Polyurethane foams RTV silicones	Homogeneous catalysis	dibutyltin diacetate dibutyltin di-octoate dibutyltin dilaurate
Esterification	Homogeneous catalysis	butanestannonic acid dibutyltin diacetate dibutyltin oxide
Glass treatment	Precursor for tin (IV) oxide films on glass	dimethyltin dichloride butyltin trichloride methyltin trichloride
Poultry management	Anthelminthic	dibutyltin dilaurate
Wood preservation	Fungicide	bis(tributyltin) oxide tributyltin naphthenate tributyltin phosphate
Agricultural chemicals	Fungicide Insecticide Miticide Antifeedant	triphenyltin acetate triphenyltin hydroxide tricyclohexyltin hydroxide bis(trineophyltin) oxide 1-tricyclohexylstannyl-1,2,4-triazole
Antifouling paints	Biocide	triphenyltin chloride triphenyltin fluoride bis(tributyltin) oxide tributyltin fluoride tributyltin chloride tributyltin acrylate polymers
Materials protection (stone, leather, paper, etc)	Fungicide Algicide Bactericide	bis(tributyltin) oxide tributyltin benzoate
Moth proofing of textiles Disinfection	Insecticide Antifeedant Bacteriostat	triphenyltin chloride triphenyltin hydroxide tributyltin benzoate

less liquids which are thermally stable to 200 °C and do not react rapidly with air or water, although, like other organotins they are essentially degraded under environmental conditions to inorganic tin compounds. Although toxic to mammals R_4Sn compounds do not possess significant biological activity and their only large commercial outlet is as precursors for other organotin compounds. Maximum biological activity occurs in the organotin series when n = 3, and triorganotins are used commercially as biocides in a number of areas.

Dialkyltin compounds R_2SnX_2, are powerful stabilisers for PVC; the most effective heat stabilisers are those containing Sn—S bonds, whereas dialkyltin bis(carboxylates) are preferred when resistance to light and weathering is required. Certain di-n-octyltin stabilisers have been approved for use in PVC for food contact applications. Surprisingly dimethyltin compounds have also shown low toxicity and dimethyltin S, S'-bis(iso-octylthioglycolate) has been approved in some countries for food contact PVC, the LD_{50} value falling within the range of that of the dioctyltin compounds. Dibutyltin compounds are used as stabilisers for non-food contact PVC and also as homogeneous catalysts in the manufacture of polyurethane foams and as cross-linking agents in room-temperature-vulcanising silicones. Mono-organotin compounds exhibit low mammalian toxicity and are thus desirable industrial compounds. To date their main use has been as synergistic additives for stabilising PVC and, to lesser extent, as esterification catalysts. There are indications [118] that they may have a place as waterproofing agents for cellulosic materials such as cotton textiles, paper and wood and as fire retardants for woollen fabrics.

Chemical Properties

The tin-carbon bond is stable to water and to atmospheric oxygen at normal temperatures and is quite stable to heat (many organotins can be distilled under reduced pressure with little decomposition). Strong acids, halogens and other electrophilic agents readily cleave the tin-carbon bond. Tin forms predominantly covalent bonds to other elements but these bonds exhibit a high degree of ionic character, with tin usually acting as the electropositive member. Triorganotin hydroxides behave not as alcohols but more like inorganic bases, although strong bases remove the proton in certain triorganotin hydroxides, since tin is amphoteric. Thus, bis(triorganotin) oxides, $(R_3Sn)_2O$, are strong bases and react with both inorganic and organic acids, forming salt-like but non-conducting and water-insoluble compounds. Tin doubly bonded to oxygen does not exist and diorganotin oxides, R_2SnO, are polymers, usually highly cross-linked via intermolecular tin-oxygen bonds. Unlike the halo-carbons, organotin halides are reactive compounds and because of their ionic character, readily enter into methathetical substitution reactions resembling the inorganic tin halides. Unlike carbon, tin shows much less tendency to form chains of tin atoms bonded to each other. Although tin-tin bonded compounds are known (for example hexa-organoditins) the tin-tin bond is easily cleaved by oxygen, halogens and acid.

Analytical Methods

When determining the environmental fate of organotin compounds problems arise from the fact that one is often dealing with extremely low concentrations. Speciation and subsequent quantitative analysis of derivatives at these very low levels present special difficulties. The determination of organotins in environmental samples has been greatly helped by the advances in determining metals at nanogram levels that have taken place in the last 6–7 years. Techniques such as high performance liquid chromatography (HPLC), gas chromatography (GC) and electrothermal atomisation (ETA) for atomic absorption spectroscopy (AAS) have all proved effective in this respect.

The analysis of organotin species in environmental samples has recently been discussed elsewhere [42]. Therefore, this section will simply outline some of the methods used to determine organotins in different media.

Aqueous Systems

By far the largest amount of published work concerned with the analysis of organotin compounds at trace levels has been on their determination in aqueous systems, either as naturally occurring species or as leachates from antifouling tests. Consequently, to review all of the publications on this topic would be outside the scope of this work. Table 2, therefore, summarises a selection of the methods that have been used for the determination of organotin compounds in water, together with their approximate limits of detection.

Table 2. Selected techniques for determining inorganic and organotin compounds in waterways [38]

Compounds investigated	Method	Approximate detection limit	Ref.
R_3SnX (R = Me, Et and Bu) R_2SnX_2 (R = Me, Et and Bu) $RSnX_3$ (R = Me, Bu and Ph)	Conversion to hydride with $NaBH_4$, collection in a trap, separation by boiling-point, detection by AAS (Atomic absorption spectroscopy)	1 ng l^{-1}	119
R_3SnX (R = Me, Et, Pr, Bu, Ph and cyclo-C_6H_{11})	HPLC (high-performance liquid chromatography) strong cation exchange columns with determination by GFAA (graphite furnace atomic absorption spectroscopy)	$5-30$ ng as Sn	129
Me_nSnH_{4-n} (n = 1–4)	Automatic purge trap sampler followed by GC (gas chromatography) separation and determination by FID (flame ionization detector)	1 μg l^{-1} on a 10 ml sample	127
Me_nSnX_{4-n} (n = 0–3)	Conversion to hydride with $NaBH_4$, separation by boiling-point and determination by FED (flame emission detection)	1 ng l^{-1} on a 100 ml sample	51
Me_nSnX_{4-n} (n = 0–3)	Extraction with benzene followed by butylation and separation by GC and determination by AAS	0.04 ng l^{-1} on a 5 l sample	76
Bu_nSnX_{4-n} (n = 1–3)	Solvent extraction, followed by methylation and separation by GC and determination by mass spectrometry	10 μg l^{-1} on a 1 l sample	157
Bu_nSnX_{4-n} (n = 1–3)	Solvent extraction, followed by pentylation and determination with a modified FPD (flame photometric detector)	10 mg l^{-1} on a 25 ml sample	151
Ph_nSnX_{4-n} (n = 0–3)	Separation by solvent extraction of the individual species followed by determination as inorganic tin by photometric methods	2 μg as Sn	104

Table 2. (continued)

Compounds investigated	Method	Approximate detection limit	Ref.
Ph_nSnX_{4-n} (n = 1–3)	Solvent extraction, followed by conversion to hydride with $LiAIH_4$, and separation and determination by GC–EC (gas chromatograph-electron capture detector)	$0.015\ \mu g\ l^{-1}$ on a 200 ml sample	203
Bu_3SnX	Solvent extraction, followed by direct photometric determination with dithizone	1 p.p.m.	79
$(Bu_3Sn)_2O$	Solvent extraction and direct photometric determination on the haematein complex	0.1 p.p.m.	174
"Organotins"	Solvent extraction, wet-ash followed by photometric determination with phenylfluorone	$1\ \mu g\ l^{-1}$ on a 200 ml sample	196
"Organotins"	Solvent extraction wet-ash followed by photometric determination with phenylfluorone	$0.01\ mg\ l^{-1}$ on a 50 ml sample	162
Bu_3SnX	Initial separation by steam distillation followed by extraction, wet-ashing and determination by inverse voltammetry	0.01 p.p.m.	241
Ph_3SnX	Concentration by evaporation, extraction and determination by inverse voltammetry as above	0.01 p.p.m.	241
Ph_3SnX	Solvent extraction and direct spectrofluorimetric determination with 3-hydroxyflavone	0.004 p.p.m. on a 50 ml sample	40
Inorganic Sn (II) and Sn (IV)	Concentration by coprecipitation and flotation. Dissolved precipitate reacted with $NaBH_4$ and determination by AAS	$0.07\ \mu g\ l^{-1}$ sample	163
Inorganic Sn (IV)	Differential-pulse and anodic stripping voltammetry direct on the sample	$4\ \mu g\ l^{-1}$ on a 25 ml sample	148
Inorganic Sn (IV)	AAS/ETA direct on the sample but with the addition of 10% ascorbic acid before the ETA (electrothermal atomization) drying stage	$50\ \mu g\ l^{-1}$	219

Soils and Sediments

Organotin compounds adsorb strongly onto soil and sediments. Consequently, although good recoveries have been obtained from 'spiked' samples, it is almost impossible to quantitatively extract these species from 'real-life' samples of soils and sediments, and no reliable methods have as yet been developed.

Air

Bis(tributyltin) oxide has been determined in air, at a level of approximately 0.1 mg m^{-3}, by Jeltes [128], using a high volume air sampler and collecting the organotin on filters which are either extracted with methyl *iso*-butyl ketone, and the extract examined by atomic absorption spectroscopy, or are extracted with toluene and subjected to gas chromatography, after in situ pyrolysis of the tributyltin compound. Zimmerli and Zimmerman [242] determined tetra, tri- and di-butyltin compounds in air by absorbing the compounds on Chromosorb 102 and then extracting with diethyl ether containing 0.3% hydrochloric acid. The organotin compounds are converted to the corresponding methyl-butyltin derivatives, by reaction with methylmagnesium chloride, separated using gas chromatography, and the tin content determined using a tin specific flame photometric detector; it is claimed that $0.05 \ \mu g \ m^{-3}$ of butyltin compounds can be determined on one cubic metre of sample. Tri, di- and mono-cyclohexyltin compounds have also been determined in air [61], with a sensitivity of approximately $0.02 \ mg \ m^{-3}$.

Biological Material

Although a large number of studies have been carried out on the effect of organotins on marine life [211], only a few have involved the actual analysis of tissue. Ward et al. determined [238] tri- and di-butyltin species in whole fish or tissues of muscle, liver, embryo and viscera, by dispersing them in concentrated hydrochloric acid, extracting the dispersion with hexane, adding tropolone and separating the organotin compounds using an alkali wash [13]; the tin content being determined by graphite furnace atomic absorption spectroscopy. Waldock and Miller [235] also used this method, together with the gas chromatography/mass spectrometry method of Meinema and his co-workers [157], to determine the organotin content of oyster tissue, and found that the two methods generally gave good agreement. Tugrul et al. [222] analysed fish, macro-algae and limpets for methyltin compounds using a method similar to Hodge et al. [119]. The determination of a range of organotin compounds in biological material has been investigated by Wada and his co-workers and is summarised in a review [234]. Aldridge et al. [7, 8] applied a spectrofluorimetric method to the determination of trimethyltin compounds in brain tissue. A fluorimetric method has also been developed by Arakawa et al. [27], using morin as the complexing agent, for the determination of dialkyltin compounds and, by inclusion of a separation step by high performance liquid chromatography, applied this to the determination of dialkyltin derivatives in various biological tissues [242].

Sources of Entry into the Environment

Organotin compounds are used in a wide range of industries [43, 98] but from the point of view of their impact on the environment these applications can be divided into those which are biocidal and those which are not. Biocidal uses for the most part involve triorganotins, and although biocides only account for about 30% of total organotin consumption, the main environmental effects are likely to come from this sector. Diorganotins and mono-organotins are used mainly in the plastics industry. Figure 3 shows schematically the routes by which organotins might conceivably enter the environment, associated with their commercial usage.

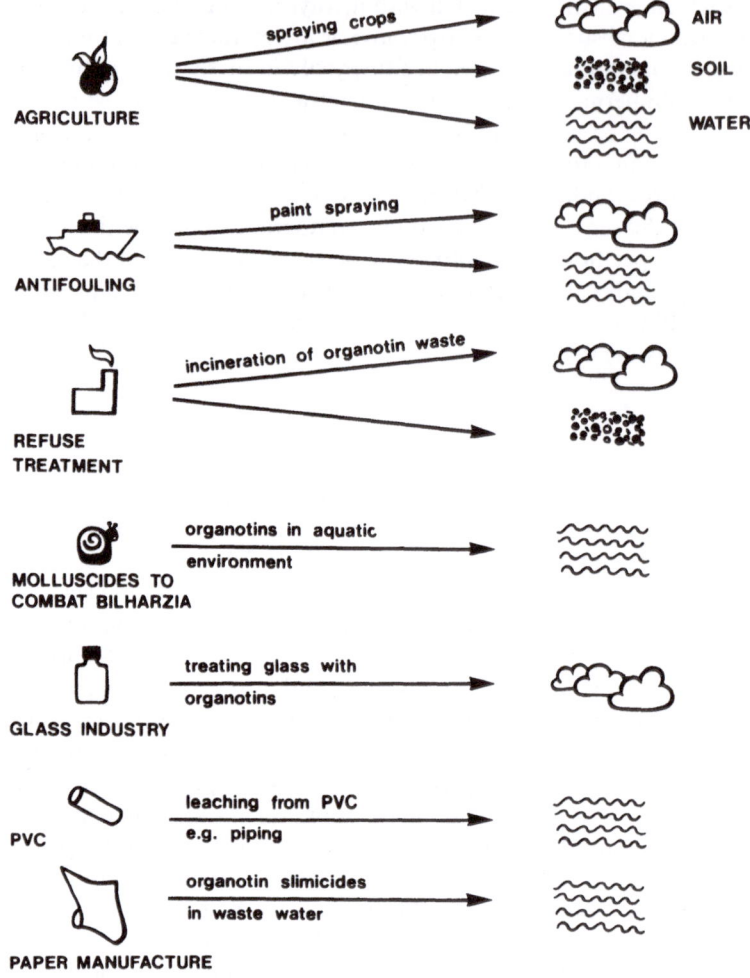

Fig. 3. Possible sources of entry of organotin compounds into the environment.

PVC Stabilisation

The major use for organotins is in PVC stabilisation where concentrations of 0.5–2.0 % of the organotin compound are effective. The stabiliser additive may be incorporated into the PVC formulation by the supplier of the polymer, along with lubricants, impact modifiers, etc. or it may be added before processing by the PVC fabricator, eg, the blow moulder who produces PVC bottles. Routes by which these organotin stabilisers might enter the environment include: emissions during processing, (eg, calendering or extruding), leaching from PVC products (eg pipes or bottles) and disposal of used PVC (incineration or land burial). The US National Institute of Occupational Safety and Health (NIOSH) recommends a threshold limiting value (TLV) of 0.1 mg/m³ air (calculated as tin) in workplace air [19]. In fact, ever since the concern about possible release of toxic vinyl chloride monomer during PVC processing,

Table 3. Tin concentrations in foodstuffs after storage for 2 months in PVC bottles at 30 °C [70]

Food product	Tin content at beginning of experiment (mg/kg)	Tin content after ageing in PVC bottle (mg/kg)	Tin extracted from bottle (mg/kg)	Organotin stabilizer extracted from PVC bottle (mg/kg)
mineral water	0.076	0.088	0.01	0.063
tomato juice	0.03	0.03	0	0
peanut oil	0.05	0.06	0.01	0.063
vegetable oil	0.08	0.09	0.01	0.063
apple juice	0	0.02	0.02	0.126
cherry soda	0	0.07	0.07	0.443
beer	0	0.01	0.01	0.063
milk[a]	0.02	0.04	0.02	0.126
red wine	0	0	0	0
blended whisky	0.01	0.02	0.01	0.063

[a] Milk sample in PVC bottle aged for 2 weeks at 65 °C

extraction systems have been rigorously upgraded and this TLV is unlikely to be reached. There have been many studies of the rate of leaching of organotins from stabilised PVC and this subject has been reviewed [193]. Rates depend on the length of the alkyl chain in the stabiliser, on the leaching medium (eg pH) and on the formulation (eg amount of plasticiser present). The migration of tin from PVC bottles into liquid foods has been studied by Carr [70]. The increase in tin content of various products after storage in PVC bottles for two months at 30 °C ranged from 0 to 0.07 mg/kg (Table 3).

The low levels of migration which have been encountered suggest that this source of environmental pollution is not significant.

When waste PVC is incinerated, any organotin compounds would be converted to inorganic forms of tin and in any case only very small amounts would be released to the atmosphere. By far the largest proportion of PVC in refuse is disposed of by land filling. Although in theory organotins could be leached from this material, eg by soil waters, migration rates are extremely low and since the PVC would be fairly uniformly distributed in the refuse with an overall low concentration, this is not likely to constitute a significant source of pollution.

Wood Preservation

Tributyltin compounds are used to protect wooden joinery and structural timbers against fungal and insect attack and are applied to wood as a 1–3% solution in an organic solvent such as kerosene. Although for remedial work the preservative may be applied by brushing or spraying, by far the biggest use is for pretreating wood in specially designed impregnation chambers where alternating vacuum and pressure cycles are applied. Such closed-system treatments are unlikely to result in much outside contamination. Once the solvent has evaporated, the wood contains the tributyltin compound securely held within its structure. Leaching of organotins from treated wood is unlikely and recent studies have shown [187] that organotin-treated

Table 4. Vapour pressure data on organotin compounds

Compound	Vapour Pressure (mm Hg)	Ref.
Bu_3SnOAc	2.7×10^{-3} (20 °C)	220
$Bu_3SnO.CO.Ph$	1.5×10^{-6} (20 °C)	24
$(Bu_3Sn)_2O$	7.5×10^{-6} (20 °C)	24
$(Bu_3Sn)_2O$	1.2×10^{-4} (20 °C)	220
$(Bu_3Sn)_2O$	1.1×10^{-5} (25 °C)	21
$(Bu_3Sn)_2O$	6.4×10^{-7} (20 °C)	150
$1\text{-(cyclo-}C_6H_{11})_3$ $Sn(1,2,4\text{-triazole})$	$< 5 \times 10^{-5}$ (25 °C)	134
Ph_4Sn	9.2×10^{-11} (25 °C)	140
Ph_4Sn	1.3×10^{-11} (25 °C)	131
Ph_4Sn	1.2×10^{-11} (25 °C)	131
Ph_6Sn_2	3.2×10^{-15} (25 °C)	131
Ph_6Sn_2	2.7×10^{-15} (25 °C)	131

wood can safely be used for interior applications since the vapour pressure of tri-butyltin compounds is low. Vapour pressures of some organotin compounds are shown in Table 4.

Agricultural Uses

In the case of agricultural applications, non-systemic organotin compounds are applied by spraying of crops; in a few cases, aerial spraying may be adopted. There thus exists the possibility of residues on crops, contamination of surrounding soil and run-off into water.

There are five organotin compounds in commercial use as pesticides; these are the fungicides triphenyltin acetate and triphenyltin hydroxide and the miticides tri-cyclohexyltin hydroxide, bis[tris(2 methyl-2 phenylpropyl) tin] oxide and 1-tricyclo-hexylstannyl-1,2,4-triazole. A detailed review of triphenyltin compounds covering their physical, chemical and biological properties, toxicology, analysis and environmental behaviour has been produced by Bock [48]. Following lengthy chronic toxicity testing, the World Health Organisation pronounced the triphenyltin compounds (triphenyltin acetate, hydroxide, chloride) as 'safe agricultural chemicals' and fixed an acceptable daily intake for man at 0–0.0005 mg/kg body weight [15]. Table 5 shows recommended dosages and waiting times (between treatment and harvesting) for the triphenyltin compounds applied to combat various crop diseases.

Tricyclohexyltin hydroxide was the first organotin compound to be developed as a miticide, following publication of a patent [132] in 1966 which claimed tricyclo-hexyltin compounds as pesticides for controlling arachnids. The hydroxide was adopted for commercial formulations and these were introduced under the name "Plictran" in 1970 [95]. Formulations based on this compound give control of motile forms of plant-feeding mites with little effect on most predacious mites and insects at recommended rates of application. The World Health Organisation has established the acceptable daily intake for this compound by man at 0–0.007 mg/kg body weight [18]. Bis[tris(2-methyl-2-phenylpropyl)tin] oxide was introduced in 1975 by Shell

Table 5. W.H.O. Recommendations for use of triphenyltin compounds to protect crops [15]

Crop	Disease	Recommended dosage kg a.i./ha	No of applications	Safety period, day	Recommended tolerance ppm
potatoes	*Phytophthora infestans* of foliage and tubers, and *Alternaria solari*	0.18—0.4	1—10	7	0.1
sugar-beet	*Cercospora beticola*	0.18—0.4	1—3	14	0.2
celery and celeriac	*Septoria apii*	0.2—0.32 (in 600—1000 litre water)	1—7	21	celery 1 celeriac 0.1
carrots	*Alternaria porri*	0.2—0.32 (in 600—1000 litre water)	1—7	7	0.2
pecans	*Fusicladium effusum, Gnomonia sp. and Mycospherella corvigen*	0.22—0.45	2—8		
rice	*various algae*	0.75—1.0 (in 1000 litre water)			
groundnuts	*Cercospora sp.*	0.2—0.6			0.05
coffee	*Colletotrichum coffeanum*	0.2—0.6 (in > 1000 litre water)		7	
cocoa	*Phytophthora palmivora*	0.2—0.6 (in 1000 litre water)			

Table 6. Reduction of triphenyltin acetate on potato leaves [80]

Days after application	Residue of triphenyltin acetate on potato leaves ($\mu g/cm^2$)	
	Application of 0.34 kg/ha	Application of 0.34 kg/ha + 6 × 0.36 kg/ha
1	0.70	0.92
2	0.40	0.69
3	0.34	0.56
4	0.29	0.51
5	0.23	0.45
6	0.21	0.40
7	0.20	0.37
8	0.19	0.34
9	0.18	0.30
10	0.18	0.29

under the trade names 'Vendex' in the USA and 'Torque' outside that country [96]. The World Health Organisation acceptable daily intake for man is 0–0.3 mg/kg body weight [20]. The most recent organotin miticide is 1-tricyclohexylstannyl-1,2,4-triazole which is marketed by Bayer AG under the trade name 'Peropal'. Toxicity assessments have shown Peropal to be acceptable for commercial use as an acaricide [112].

The concentration of triphenyltin compounds on treated plants decreases rapidly because of losses in wind and rain and degradation in the atmosphere and under the influence of light. A number of studies on this subject have been reported, for example the study by Coussement [80] (Table 6). The FAO (WHO) joint meeting report [15] included data on triphenyltin hydroxide, acetate and chloride residues in various foodstuffs such as potatoes, carrots and sugar beets. Maximum concentrations in potatoes and carrots only rarely exceeded 0.1 mg/kg. Residues in food, fruit and vegetables derived from direct contact with the compound could be considerably reduced by washing. Residues of tricyclohexyltin hydroxide on apples and pears decline by 50% in about 3 weeks due to photodegradation. A 20–50% reduction can be achieved by washing and most of the residues can be removed by peeling the fruits after which only 0.1 mg/kg (as tin) may be expected in the fruit flesh.

There is no evidence so far that triphenyltin compounds can act systemically. In experiments with radioactively labelled as well as unlabelled organotin compounds, no detectable or only extremely low concentrations were found in the tested plants [48]. Thus, once the inedible sprayed portions are stripped away and the foodstuff cleaned and cooked, there is very little risk of human consumption of organotin compounds. There still remains the risk, in theory, of animals eating contaminated leaves and organotins entering the animal destined for human foodstuffs. However, when cows were fed on sugar beet leaves containing 1 mg/kg triphenyltin acetate, only 0.004 mg/kg was found in milk [60]. In meat, no tin content above the blank value could be detected. It may be concluded that human consumption of organotin compounds in milk was only possible when the animals were slaughtered immediately after consuming treated leaves. Further, it can be assumed that any phenyltin residues still present will be partly or completely destroyed by cooking. Waiting times are always specified between crop spraying and harvesting, to allow for elimination of the organotin in agricultural practice.

The other potential source of environmental pollution is by overspray of surrounding soil during crop treatment. It is known that triorganotin compounds are strongly adsorbed on soil. Barnes et al. [31] found that ^{14}C-triphenyltin acetate could not be leached with water from a 25 cm layer of an agricultural loam over a period of 6 weeks and subsequent analysis of the soil for radioactivity showed that over 75% of the organotin compound remained in the top 4 cm. ^{14}C-triphenyltin chloride in clay was studied by Leeuwangh et al. [141] and strong adsorption was again found, the ratio of triphenyltin chloride in sediment to that in water was about 20:1. Strong affinity of soil particles for tributyltin compounds [33, 114] and for other triorganotins [130, 199) has been demonstrated. It thus seems likely that after spray application, triorganotins will be adsorbed on soil before they can enter surface waters.

Antifouling Coatings

The use of organotin compounds in antifouling paints constitutes an area where these compounds are released directly into the aquatic environment albeit in very small, controlled, quantities. The hulls of vessels immersed in water for any length of time become subject to attachment of many aquatic organisms such as barnacles, mussels and algae. These result in a significant lowering in operating efficiency with consequently increased running costs. Providing a coating from which biocidal organotins are slowly leached out into a water layer surrounding the hull can effectively inhibit such fouling. The development and use of organotin-based antifouling coatings have been reviewed by Evans and Hill [97]. Initially these antifouling systems consisted of simple paints from which the toxicant was leached by the action of sea water. Later, more sophisticated systems, such as rubber impregnated with organotins or organotin polymer systems, offered a more controlled rate of release and a longer effective life for the coating. Self-polishing coatings relate rate of release of toxicant to ship motion, and a self-levelling effect exists in the coating layer.

From the point of view of environmental pollution, the greatest threat exists when vessels are stationary in harbours. A significant build-up of organotins can occur in the water of harbours and marinas containing large numbers of small pleasure craft, mainly painted with older, conventional paint systems. In freshwater marinas levels of bis(tributyltin) oxide as high as 539 ng/g have been found in the top 2 cm layer of sediment [149]; in general much lower levels are found in sediments of lakes and rivers [110]. In English coastal waters, concentrations of bis(tributyltin) oxide have ranged from 'not detectable' to 0.43 µg/litre [235]. Sediment concentrations of various organotin compounds, determined in different parts of the world, ranged from 'not detectable' to 12.3 ng/g [188, 222]. Hall and Pinkney [110] have surveyed acute and sublethal effects of organotins in aquatic biota; they conclude that concentrations of organotin compounds in most aquatic habitats (lakes and rivers) would be likely to have minimal short-term effects on aquatic biota. Long-term effects are, however, possible, depending on the species and life stage. The most serious environmental impact would occur in and around harbour or marina areas; the upper surface microlayer in crowded harbour areas would be a toxic habitat for most aquatic species. Lower depths in the harbours could also contain sufficient concentrations of organotin compounds to cause environmental problems with the aquatic biota of the area. Recently the use of organotin paints on small pleasure craft has been circumscribed in parts of Europe, following reports that build-up of organotin compounds was affecting oysters grown on nearby oyster farms. Specifically, triorganotin compounds at very low levels have been shown to cause reduced growth and shell deformity in the commercially cultivated Pacific oyster, *Crassotrea gigas* Waldock and Thain reported [237] that *Crassotrea gigas* spat, exposed to bis(tributyltin) oxide concentrations of 0.0002 and 0.002 mg/litre for 56 days, reduced growth and increased shell thickening. Alzieu et al. described [10] malformations of the shell of the Pacific oyster after exposure to tributyltin concentrations of approximately 0.0002 mg/litre for 110 days. These anomalies were characterised by the hypersecretion of a gel which was later enclosed by a fine calcium layer, thereby forming gelatinous pockets. In some cases this was followed by shell thickening, yielding a leafy appearance. Shellfish can bioconcentrate organotins from both water and from suspended material. Thus Waldock et al. reported bioconcentration factors (BCFs) of 2300 to 11,400 for *Crassotrea gigas* exposed to bis(tributyltin) oxide (0.0002 or 0.002 mg/litre) for 56 days [237], and BCFs of 2000 and 6000 for *Crassotrea gigas* exposed to bis(tributyltin) oxide (0.002 and 0.001 mg/litre) for about 21 days [236]. Oysters exposed to bis(tributyltin) oxide were shown to rapidly reach an equilibrium plateau and slowly depurate after an initial dose. A change to polymer-bound organotin antifouling coatings is likely to reduce levels of organotin compounds which are released under stationary conditions and current legislation in the UK is encouraging this trend [25]. According to this legislation, from January 1986, controls will be imposed to prevent the supply of copolymer antifouling paints containing more than 7.5% of organotin (measured as tin in the dry paint film) and all other antifouling paints containing more than 2.5% organotin [26].

Other possible means of ingress of organotins into the environment result from application of organotin-based antifouling coatings to the hulls of ships and removal of worn coatings in dry dock. Adema and Schatzberg [2] have examined this problem. Paints are applied using airless sprayers and a paint transfer efficiency of 91–95% can be expected. (This is the ratio of weight of paint on the hull to sum of this weight plus the weight of paint overspray). It was estimated that at least 99% of the paint could be prevented from entering the harbour by the use of mobile sweeping and vacuuming equipment. Paint removal is by abrasive blasting, preferably with a water cone to suppress the spreading of dust. A future development is likely to be the use of cavitating water jets contained within a shroud. Disposal of spent abrasive containing organotin paint chips is now permitted by the US Environmental Protection Agency in properly managed sanitary landfills. Leaching tests on contaminated grits in contact with soil showed that organotin released from the abrasive was effectively retained on the soils [114]. A rotary kiln incinerator was also evaluated as a means of thermally detoxifying organotin-containing waste [1]; the process was shown to be effective, but expensive. The feasibility of detoxification of organotin-contaminated wastes by treatment with nitric acid has been demonstrated in the laboratory [91].

Molluscicides

World Health Organisation approval of the use of bis(tributyltin) oxide for control of the snail vectors of Schistosomiasis is pending the result of long-term carcinogenicity tests which are in progress with this chemical. Schistosomiasis is believed to affect between 200 and 300 million people in Central and South America, Africa and Asia. Schistosomes are parasitic blood flukes and in the miracidial larvae forms they enter specific species of aquatic snail before emerging as free-swimming cerceriae at which state they can infect humans. Control methods have been proposed in which snails are controlled by the use of small floating rubber pellets which release very small amounts of organotin toxicant [66–68]. Since snail destruction is achieved through chronic intoxication, only ultra-low concentrations of organotin are necessary and environmental impact should be small. Extensive tests have taken place in many parts of the world to confirm that no deleterious effects on fish or other aquatic life occur at the concentrations used [93].

Other Uses

Diorganotin compounds are often contained in polyurethane foams, at levels of about 0.02% w/w, since they catalyse the formation of these foams, particularly the flexible varieties. The methods of release of these organotins into the environment will be the same as for PVC stabilisers, *ie* land-fill and incineration, and again will not result in much environmental contamination. Although flammability is a problem with these foams, and toxic products are likely to be released during combustion, any organotin compounds present would be decomposed at these temperatures. Low-tonnage uses of organotin compounds are varied and include disinfectants, glass-strengthening treatments and anthelmintic treatments for poultry; these are not expected to contribute significantly to the environmental burden, due to the low tonnages involved.

Environmental Interactions

The principal routes by which organotin compounds are introduced into the environment have been discussed; the degree to which they constitute a long-term hazard depends on the extent to which they become established in this environment and in particular their possible entry into food chains. Amounts of organotins found in the atmosphere are negligible so that interactions from this source are unlikely to be significant. Also, organotins are strongly adsorbed on soils and sediments, so that soil/water interactions are limited. A number of studies have been conducted to ascertain the long-term effect of organotin compounds on soil fertility. Bollen and Tu examined [50] the effect of bis(tributyltin) oxide (10 and 100 ppm) on microbial activity relevant to soil fertility. It was shown that at these concentrations the triorganotin had no biologically significant impact on soil microbial populations, ammonification, nitrification, sulphur oxidation or soil respiration. The interaction of the methyltin chlorides, Me_nSnCl_{4-n}, with humates, peat and related materials has been studied, using the tin-113 labelled compound [167]. Triorganotins do not act as systemic biocides in crop protection, so that uptake by plants from contaminated soil is not likely to be significant.

Thus, the most important area to be considered is that of organotins in the aquatic environment. Tin has been reported [155, 233] to be naturally present in sea water up to 3 mg/litre but there have been few reports of its occurrence in marine algae, plankton, bacteria, flowering plants, protozoa, crustaceae or fish. Organotin compounds tend to concentrate in aquatic biota and would thus pose a risk to the food chain if they passed from phytoplankton to the higher trophic levels. Evans and Laughlin [99] exposed *Artemia* to water containing 0.020 mg/litre bis(tributyltin) oxide: concentrations of 6.2 ± 1.1 mg bis(tributyltin) oxide per g *Artemia* (wet wt) were subsequently found. When these *Artemia* were used as a food source for the mud crab *Rhithropanopeus harrisii* accumulation of bis(tributyltin) oxide was much higher than when crabs were only exposed to water containing the organotin. The ability of shellfish to bioconcentrate organotins from sediments and suspended particles has already been discussed.

Aqueous Chemistry of Organotins

By far the largest number of studies of the nature of organotin compounds in aqueous solution has been conducted on methyltin derivatives, since other organotins frequently do not have a sufficiently high solubility in water to permit spectroscopic investigation (Table 7).

Table 7. Aqueous solubilities of tri- and di-organotin compounds at room temperature[a]

Compound	Solubility (ppm)	Ref.
Et_3SnOH	35,000	116
Et_3SnOAc	7,500	232
Pr_3SnCl	50	232
$(Pr_3Sn)_2O$	50	232
iso–Pr_3SnCl	25	232
Bu_3SnF	6	34
Bu_3SnF	4.5	169
Bu_3SnCl	50	232
Bu_3SnCl	17	73
Bu_3SnCl	16	73
Bu_3SnCl	5.4	169
Bu_3SnOAc	256	79
Bu_3SnOAc	65	73
Bu_3SnOAc	50	232
Bu_3SnOAc	16	73
Bu_3SnOAc	6.4	169
Bu_3SnOAc	5	79
$(Bu_3Sn)_2O$	19.5	79
$(Bu_3Sn)_2O$	18	73
$(Bu_3Sn)_2O$	8	73
$(Bu_3Sn)_2O$	8–10	157
$(Bu_3Sn)_2O$	3	232
$(Bu_3Sn)_2O$	1.4	79
$(Bu_3Sn)_2S$	1	232
Bu_3SnSO_3Me	31,000	45
Bu_3SnSO_3Et	29,000	45
Ph_3SnF	1.2	34
Ph_3SnCl	5.2	169
Ph_3SnCl	1	232

Table 7. (continued)

Compound	Solubility (ppm)	Ref.
Ph₃SnOH	1.2	34
Ph₃SnOH	1.2	204
Ph₃SnOH	1	232
Ph₃SnOAc	3.3	31
Ph₃SnOAc	2.9	158
Ph₃SnOAc	1	232
(Ph₃Sn)₂O	1	232
Ph₃SnSO₃Et	1,000	45
Ph₃SnSO₃Ph	1,000	45
Oct₃SnCl	1	232
$(\text{cyclo-}C_6H_{11})_3\text{Sn}-\overset{\displaystyle \lceil\qquad\quad\rceil}{\underset{\displaystyle \text{H}\quad\ \text{H}}{\text{N.C:N:C:N}}}$	1	112
[(PhMe₂CCH₂)₃Sn]₂O	0.005	16
Me₂SnCl₂	20,000	232
Bu₂SnCl₂	92	73
Bu₂SnCl₂	50	232
Bu₂SnCl₂	6–8	157
Bu₂SnCl₂	4	73
Bu₂Sn(OAc)₂	6	232
Bu₂Sn(O.CO.ⁿC₁₁H₂₃)₂	3	232
Bu₂Sn(O.CO.ⁱOct)₂	6	232
[Bu₂SnO.CO.CH:CH.CO.O-]ₙ	6	232
Oct₂SnCl₂	1	232
Ph₂SnCl₂	50	48

^{a} Solubility values marked refer to distilled water: unmarked refer to sea-water

Tobias demonstrated [218] that trimethyltin compounds in aqueous solution at pH < 5 exist primarily as the trimethyltin cation, Me_3Sn^+, and at higher pH as Me_3SnOH. The trimethyltin cation has been shown [55] to be hydrated and to have a trigonal bipyramidal tin atom geometry with axial water molecules (A; R = Me).

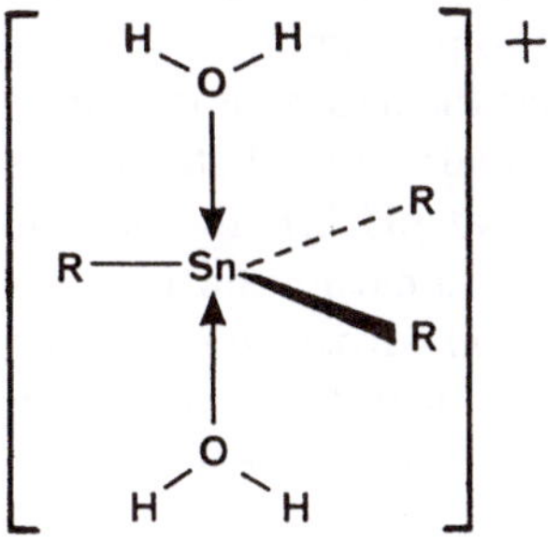

(A)

A ^{119}Sn NMR investigation of 0.5 M aqueous solutions of Bu_3SnSO_3R (R = Me, Et) revealed [45] that the triorganotin species in solution was the hydrated cation,

$(Bu_3Sn(H_2O)_2)^+$, and this has been shown [88] crystallographically to possess Structure A (R = Bu). With regard to tributyltin derivatives released into the marine environment via antifouling coatings, Monaghan et al. [160] considered that these are in the form of triorganotin hydroxides. However, Guard et al. [108] proposed that in seawater tributyltin compounds exist in an equilibrium between the hydrated tributyltin cation, tributyltin chloride, bis(tributyltin) carbonate and tributyltin hydroxide.

Speciation of dimethyltin compounds in aqueous solution has also been investigated [218]. At ph < 4 the predominant species is the dimethyltin cation, Me_2Sn^{2+}, which is hydrated and has an octahedral tin atom geometry with *trans*-methyl groups (B) [146, 147].

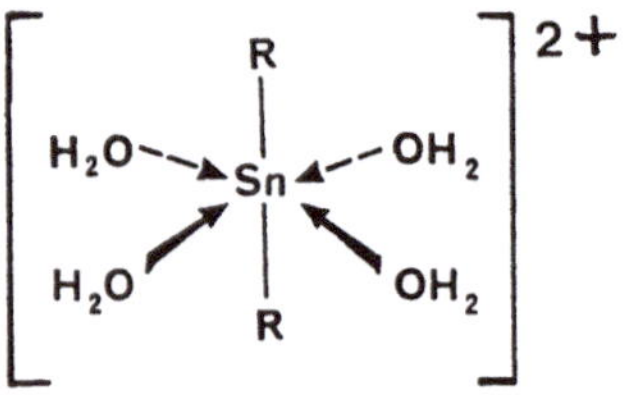

$$\left[\begin{array}{c} R \\ H_2O \text{--} \text{Sn} \text{--} OH_2 \\ H_2O \quad OH_2 \\ R \end{array} \right]^{2+}$$

(B)

However, at environmental pH (6–8), the main species is $Me_2Sn(OH)_2$ [218].

Tobias suggested [218] that mono-alkyltin derivatives exist in solution only as hydrated oxides. However, studies [46] with methyltin trichloride have shown that at pH 1–4, a concentration-dependent equilibrium exists:

$MeSn(OH)Cl_2 \cdot 2H_2O \leftrightharpoons MeSn(OH)_2Cl \cdot nH_2O \leftrightharpoons MeSn(OH)^{2+}$ at low concentration, $MeSn(OH)^{2+}$ was the main species and is presumably in equilibrium with other hydroxy derivatives:

$MeSn^{3+} \leftrightharpoons MeSn(OH)^{2+} \leftrightharpoons MeSn(OH)_2^+ \leftrightharpoons MeSn(OH)_3$

With increasing pH, the right hand species are favoured and at environmental pH, $MeSn(OH)_3$ might be expected to predominate.

It is also relevant to consider the nature of inorganic species in aqueous solution. Pettine et al. [173] studied the hydrolysis of stannous salts in seawater from 0.1 to 1.0 M at 20 °C and found that at pH 8.1, the predominant species was $Sn(OH)_2$. Other workers [164], studying stannic nitrate (8×10^{-6} g ion/litre) concluded that above pH 3, the inorganic tin (IV) hydroxy complex was primarily $Sn(OH)_4$. It would thus be expected that at environmental pH, tin compounds, R_nSnX_{4-n} (n = 0–3) would exist in aqueous solution as simple neutral hydroxides.

Methylation

The feasibility of the environmental methylation of certain inorganic heavy metal compounds has been recognised for many years [72], and concern often arises over this phenomenon due to the increased toxicity of many organometallic species in comparison with their inorganic counterparts. With respect to tin the first laboratory studies were conducted in 1974 by Huey et al. [122], using pure cultures of tin-resistant *Pseudomonas* bacteria from Chesapeake Bay. Incubation with $SnCl_4 \cdot 5 H_2O$ led to the production of what was believed to be a dimethyltin species. Interestingly, when mercury (II) compounds were added to the tin system, quantities of methylmercury (CH_3Hg^+) were produced, and the authors considered that this may have arisen due to methyl transfer from methyltin species to mercury. Brinckman and Iverson [53] later proposed a 'mercury-tin crossover' and this hypothesis is given support by the fact that Schramel et al. reported [185] that mercury and tin accumulate together in some aquatic plants in Bavarian rivers. In subsequent studies, the products of the in vitro biomethylation of inorganic tin (IV) by the *Pseudomonas* strain were methyl-stannanes, Me_nSnH_{4-n}, where n = 2–4 [127].

The formation of methyltin compounds in sediments has been described [77, 107, 111]; both $SnCl_2$ and $SnCl_4$ were found to be methylated and in the study of Hallas [111] the species formed from $SnCl_4 \cdot 5 H_2O$ were the hydrides Me_2SnH_2 and Me_3SnH (on the basis of gas chromatographic retention times and mass spectral data). Tri-methyltin chloride [77] and hydroxide [107] are converted to tetramethyltin in sediments; this process appears to be slow. Since Me_4Sn was also produced from Me_3SnOH in sterile sediments, the tetraalkyltin derivative could be formed by a chemical pathway [77], for example, by disproportionation of the initially formed bis(trimethyltin) sulphide [82].

$$3 Me_3SnSSnMe_3 \rightarrow (Me_2SnS)_3 + 3 Me_4Sn$$

It has since been established that 95% of the tetramethyltin produced from trimethyltin precursors in sediments, occurs via a chemical rather than a biological pathway [83]. A study of tricyclohexyltin hydroxide in soils under aerobic and anaerobic conditions indicated an absence of methylated products [17].

There have been a number of laboratory studies using chemical methylating agents for tin (II) and tin (IV) compounds. Methylcobalamin is the methyl coenzyme of cyanocobalamin (Vitamin B_{12}) and this has been demethylated by $SnCl_2$ in aqueous hydrochloric acid solution, in the presence of an oxidising agent $(Fe^{3+}$ or $Co^{3+})$, to form a monomethyltin species [90, 101]. The $SnCl_3^-$ ion present in the acidic solution may be first oxidised to a trichlorostannyl radical, $\dot{S}nCl_3$, which could then cleave the Co-C bond homolytically to produce $MeSnCl_3$.

Craig and Rapsomanikis have also studied [84] the reaction of $SnCl_2$ with methylcobalamin in aqueous solution at pH 1 under aerobic conditions and found the products formed were a mixture of $MeSn(OH)_2 \cdot 2 H_2O$, $MeSn(OH)_2 \cdot nH_2O$ and $[MeSn(OH) (H_2O)_4]^{2+}$.

Thayer found [214, 215] that finely divided SnO_2 reacts with methylcobalamin in aqueous hydrochloric acid solution to form methyltin derivatives. The reaction was very slow, but this may have been due to the particle size of the SnO_2 [216] since when a solution of $SnCl_4 \cdot 5 H_2O$ dissolved in nitric acid is slowly neutralised, a very finely divided precipitate is gradually formed and this reacts more rapidly with methylcobalamin. When liquid $SnCl_4$ was added to aqueous methylcobalamin, an even more rapid reaction occurred, producing methylstannanes, Me_nSnH_{4-n} (n = 1, 2), with $MeSnH_3$ predominating [216]. Similar products were formed when tin-containing sediments were treated with methylcobalamin [216]. The reactions of di- and tri-methyltin chlorides with methylcobalamin in aqueous solution have been studied, and although tetramethyltin was detected as a reaction product, it was believed that dismutation was the major route of production [84], *ie*

$$4\,(CH_3)_2Sn^{2+} \leftrightarrows 2\,(CH_3)_3Sn^+ + 2\,CH_3^+ + 2\,Sn^{2+}$$
$$2\,(CH_3)_3Sn^+ \leftrightarrows (CH_3)_4Sn + (CH_3)_2Sn^{2+}$$

Chau et al. [77] found that methyl iodide, CH_3I, produced by certain algae and seaweeds [81], could methylate inorganic tin (II) salts in an aqueous medium, to form monomethyltin species, along with small amounts of di- and tri-methyltin derivatives, whereas tin (IV) compounds did not react. More recently, Brinckman and his co-workers [153] demonstrated the oxidative methylation of SnS by methyl iodide in aqueous solution, under mild aerobic or anaerobic conditions to form $MeSnI_3$. The methylation of inorganic tin (II) and methyltin (IV) ions by CH_3I and $(CH_3)_2Co(N_4)^+$ in the presence of the oxidising agent MnO_2 has been studied [178], and recently the feasibility of the methylation of tin (II) compounds under simulated environmental conditions was demonstrated [85]. It has also been found [217] that, following treatment of naturally occurring sediments with CH_3I, dimethyltin dihydride and di-methyltin sulphide were formed almost immediately, whilst tetramethyltin was detected after standing overnight. An additional observation, reported by Brinckman et al. [54], was that methyl iodide undergoes reaction with tin (IV) in its common oxide ore, cassiterite, to solubilise tin, although no methyltins are found.

From these studies it may be concluded that both inorganic tin (II) and tin (IV) compounds and methyltin (IV) derivatives may be either biologically or chemically methylated under simulated enviromental conditions. With the increasing development of analytical techniques for determining very low concentrations of organotins, a number of workers [51, 62, 76, 119, 221, 222] have reported concentrations of methyl-tin species in various media, such as seawater, lake water, rain water, tap water, human urine, sediments and shell samples at the p.p.b. level. It is possible that some of the methyltins detected could have arisen from the use of di- and mono-methyltin derivatives as stabilisers for PVC, although the presence of these compounds in open seawaters and unpolluted rivers and estuaries is more likely to be due to enviromental methylation processes.

Far less evidence has so far been obtained to support the possibility of the methyla-tion of different organotin derivatives, and Blair et al. [38] were unable to detect Bu_3SnMe in cultures of tributyltin-resistant bacteria which had been inoculated with tributyltin chloride. However, Maguire [149] has detected Bu_3SnMe and Bu_2SnMe_2, compounds which have no anthropogenic origin, in the sediment of some harbours which also contained tributyltin species. The implications of the environmental methylation of tin compounds are difficult to assess; however, the levels so far detected do not justify concern. As will be shown, organotins in the environment may be degraded by a number of routes and a steady-state condition may exist, whereby entry of methyltin compounds into the environment may be balanced by their removal through degradation.

Degradation Mechanisms of Organotins

An advantage of organotin compounds in many biocidal applications is the fact that they are subject to degradation under environmental influences. This degradation may be defined as the progressive removal of the organic groups linked to the tin atom, according to the scheme:

$$R_4Sn \rightarrow R_3SnX \rightarrow R_2SnX_2 \rightarrow RSnX_3 \rightarrow SnX_4$$

Such a stepwise loss of organic groups is accompanied by a progressive lowering in biological activity, so that once an organotin has exerted its biocidal function, it poses no long-term hazard to its surroundings and this is especially beneficial when such compounds are used for crop protection. Early studies [59, 60, 71, 113] emphasised the 'disappearance' of the compounds, but since then, more detailed, sophisticated research has endeavoured to identify the stages and products of degradation.

Three principal influences can be defined as promoting degradation; these are ultraviolet irradiation, biological cleavage and chemical cleavage. However, the subject of chemical cleavage will not be discussed in this work owing to the immense number of reactions that can result in the breaking of a tin-carbon bond; such reviews are available elsewhere [89, 165, 176]. Two further processes are capable of inducing degradation, but are not likely to be of environmental significance; these are thermal cleavage (since temperatures above 200 °C are needed to break the Sn-C bond) and gamma irradiation (owing to its negligible intensity at the earth's surface). The effects of gamma irradiation should, however, be briefly considered since this form of energy has a possible application in the sterilisation of food, some of which may be packaged in organotin stabilised PVC. Allen and his co-workers [9, 56] have recently shown that gamma irradiation of PVC containing diorganotin stabilisers, causes the organotin to degrade to tin (IV) chloride as follows:

$$R_2Sn_2X_2 \xrightarrow[HCl]{\gamma} R_2SnXCl \xrightarrow[HCl]{\gamma} R_2SnCl_2 \xrightarrow[HCl]{\gamma} RSnCl_3 \xrightarrow[HCl]{\gamma} SnCl_4$$

UV Irradiation

The atmosphere plays a large part in altering the intensity and wavelength of light reaching the earth's surface. Due to a thin band of ozone almost all of the sun's emitted radiation in the UV region below 290 nm is absorbed, although a long-term effect of the lower wavelength of light may be observed [239]. However, light of 290 nm wavelength possesses an energy of approximately 300 kJ mol^{-1}, and this is above the typical energy (190–200 kJ mol^{-1}) [198] required to break a Sn-C bond. Therefore, provided absorption of light takes place, degradation can occur.

The UV degradation of tricyclohexyltin hydroxide has been studied by Getzenda-ner and Corbin and it was suggested [106] that dicyclohexyltin, monocyclohexyltin and inorganic tin species were produced. Akagi and Sakagami [3] reported that, under their experimental conditions, UV irradiation of a solution of triphenyltin chloride yielded a mixture of triphenyltin, diphenyltin and phenyltin as well as inorganic tin compounds within 6 hours. Similarly, irradiation of dialkyltin derivatives resulted in a mixture of di- and mono-alkyltin compounds and inorganic tin, also within a period of 6 hours. Total degradation to inorganic tin required irradiation for more than 400 hours. Chapman and Price [74] irradiated triphenyltin acetate on a watchglass with light of wavelengths greater than 235 nm and greater than 350 nm; diphenyltin, monophenyltin and inorganic tin species were identified and quantitatively deter-mined. Degradation rates were greater at the lower wavelengths, confirming the findings of other workers [156]. Similar patterns were observed when bis(tributyltin) oxide and tributyltin acetate were irradiated [75]. Klötzer and Thust [137] studied the decomposition of bis(tributyltin) oxide on various matrices under a range of tem-peratures and irradiation intensities. In further work it was shown [136] that the relative rates of breakdown of tributyltin chloride and bis(tributyltin) oxide were similar; however, dibutyltin dichloride degraded faster than dibutyltin oxide. These studies were only concerned with the decomposition of the starting compounds and did not attempt to identify breakdown products.

The low aqueous solubility of many organotin compounds makes degradation studies in water particularly difficult due to the analytical problems involved. However, Soderquist and Crosby reported [204] the action of UV light on triphenyltin hydroxide in water and showed that under simulated environmental conditions a diphenyltin species was produced which underwent further breakdown, possibly forming a polymeric monophenyltin derivative $(PhSnO_xH_y)_n$. It was not, however, demonstrated whether or not this polymeric species would eventually degrade to inorganic tin.

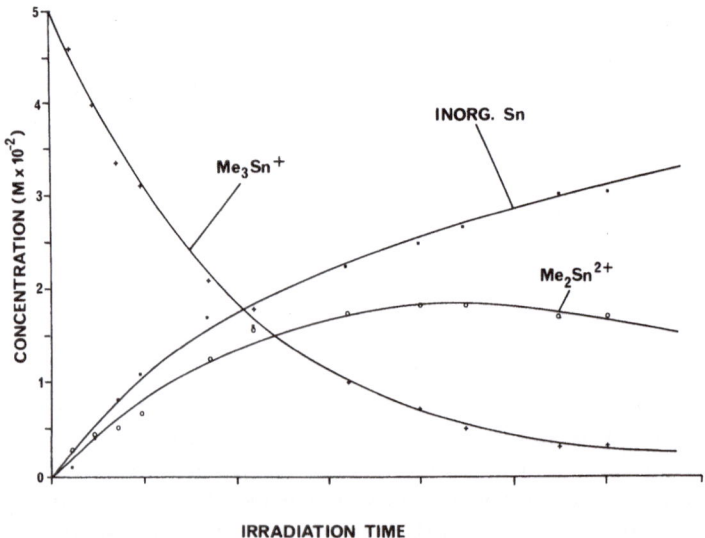

Fig. 4. The UV degradation of the trimethyltin cation, $(Me_3Sn(H_2O)_2)^+$, in water

The gradual disappearance of triphenyltin acetate in freshwater, seawater and sewage water has been noted by Odeyemi and Ajulo [166] but the breakdown products were not identified. Maguire et al. [150] studied the UV degradation of tributyltin species in water and found that the process involved sequential debutylation through to inorganic tin. A similar reaction was shown to occur slowly in natural sunlight. Sherman has studied [197] the degradation of tributyltin fluoride and chloride at 200 ppb in aqueous media (distilled water and various buffer solutions). He concluded that degradation progressed through hydrolysis of the halide bond to form a tributyltin hydroxy species which then dehydrates to tributyltin oxide and, subsequently, degrades to dibutyltin oxide. The halides and degradation products were estimated to have a life of about 60 days in aqueous media. Blunden [39] has studied the UV degradation of methyltin chlorides, Me_nSnCl_{4-n}, in aqueous solution and established the approximate relative rates of breakdown of the methyltin species. The final product was shown to be hydrated tin (IV) oxide (Fig. 4).

Biological Cleavage

The question of biodegradation of organotin compounds is particularly important in situations where, for example, the compounds are not directly exposed to light, such as in soil or on the sea bed. There is considerable evidence that under certain conditions, micro-organisms will degrade these compounds. Barnes et al. showed [31] that carbon-14 labelled triphenyltin acetate in soil could be broken down to inorganic tin and since carbon dioxide was evolved, and as similar breakdown did not occur in sterile soil, this degradation way ascribed to the ability of certain micro-organisms to metabolise the organotin compound. A similar result was obtained by Suess and Eben [208]. Barug and Vonk [33] have shown, using carbon-14 labelled bis(tributyltin) oxide, that it can be degraded in soil by microorganisms. Other work has shown this compound can be dealkylated by Gram-negative bacteria (*Pseudomonas aerugenosa* [32] and *Alcaligenes faecalis* [32]) and by the Green Alga, *Ankistrodesmus faecalis* [152]. In contrast, certain other Gram-negative bacteria have been found [32] to have no influence on the degradation of bis(tributyltin) oxide and Blair et al. [37] have recently demonstrated the exclusive uptake of the tributyltin cation $(Bu_3Sn(H_2O)_2)^+$ by a number of Gram-negative tin-resistant heterotropes, without degradation. Sheldon [195] reported the breakdown of carbon-14 labelled bis(tributyltin) oxide, tributyltin fluoride and triphenyltin chloride in soil and showed that degradation occurred faster under aerobic than under anaerobic conditions.

Bruggemann and co-workers [59, 60] reported that triphenyltin acetate on sugar beet leaves was rapidly broken down during the silage process. Within 5 weeks the concentration of the organotin decreased from 2410 mg/kg of fresh leaves, to zero. It was not established, however, whether degradation was due to microbial action or resulted from the low pH (3.5–4.0) of the silage process. Biodegradation of bis(tributyltin) oxide to inorganic tin by biological sewage treatment has been studied by Stein and Kuster [205].

Other examples of degradation of organotin compounds by micro-organisms have been cited in the area of wood preservation. Bis(tributyltin) oxide has been shown to be degraded by various individual microfungi such as *Coniophora puteana*

[32, 92], *Coriolus versicolor* [32, 115], *Chaetomium globosum* [32], *Aureobasidium pullulans* [69, 92] and various *Phialophora* species [209], as well as fungal culture filtrates of *Coniophora puteana* [209], *Coriolus versicolor* [209], and *Sistotrema brinkmannii* [168]. There is, however, the additional possibility that organotin degradation in wood is due to acid cleavage, because both formic and acetic acids may be present in some types of timber. In virtually all of the studies carried out to date, only di- and mono-butyltin species have positively been identified as breakdown products in wood. The failure to detect inorganic tin may be due to the problems associated with extracting the total tin content from the wood sample. Moreover, it has been observed [117] that as the degradation of bis(tributyltin) oxide in wood proceeds, so the amount of an unextractable tin residue increases with time.

Conclusions

The pattern of evidence which has been accumulated suggests that within a generally consistent pattern of behaviour, organotins will degrade in natural media and this has been demonstrated for triphenyltin [105, 161], tributyltin [129] and tricyclohexyltin [106] compounds.

The effect of the X group on the rate of degradation of an R_nSnX_{4-n} compound is not known, but at the levels that organotins are likely to be present in the environment, they will probably either exist, or be rapidly converted to oxides or hydroxides [31], carbonates [44, 202], or hydrated cations [45, 55, 218].

Little mention has been made of the timescales involved in the degradation processes, since although some of the studies have reported half-lives, they relate only to the specific laboratory experimental conditions and so are not directly comparable. Additionally, it can be misleading to relate experimental timescales to an environmental situation where the actual rate of breakdown will be dependent on many factors, eg intensity of sunlight, concentration of suspended matter in waterways. However, Freitag and Bock [105] have estimated half-lives for triphenyltins under environmental conditions to be between a few days and 140 days and Maguire et al. [150] estimated the environmental half-life of bis(tributyltin) oxide in water to be approximately 89 days.

Toxicology of Organotin Compounds

Toxicity

The literature on the toxicity of organotin compounds is extensive, but it is difficult to compare the findings because of the great differences in the design of the experiments. These differences relate to the substances studied, their purity and form of dosage (undiluted, concentration, dissolved, suspended, etc), the mode of administration (oral, percutaneous, intraperitoneal, intravenous, etc) and the species used (mouse, rat, rabbit, guinea pig, cat, dog, etc). In addition, there are differences in the number of administrations, number of animals employed per experiment and the period of observation. One of the most comprehensive reviews of toxicity and toxicolo-

Table 8. Acute oral LD_{50} values for some organotin compounds [200]

Compound	LD_{50}^a mg/kg	Test animal m = mouse r = rat
Tetrabutyltin	> 4000	r
Bis(tributyltin) oxide	148–234	r
Tributyltin fluoride	200	r
Tributyltin acetate	125–380.2	r
Tripropyltin oxide	120	r
Triphenyltin fluoride	486	m
Triphenyltin chloride	118–135	r
Triphenyltin hydroxide	108–360	r
Triphenyltin acetate	125–491	r
Tricyclohexyltin hydroxide	235–540	r
Fenbutatin oxide	2630	r
l-tricyclohexylstannyl-1,2,4-triazole	631	r
Dioctyltin-S,S' -bis (isooctylthioglycolate)	1200–2100	r
Dioctyltin maleate	4500	r
Dibutyltin isooctylthioglycolate	500–1037	r
Dibutyltin oxide	487–520	r
Dibutyltin diacetate	109.7	m
Dibutyltin dilaurate	175–1600	r
Methyltin trichloride	575–1370	r
Methyltin tri(isooctylthioglycolate)	920–1700	r
Ethyltin trichloride	200 (LD_{100})	r
Butyltin trichloride	2200	r
Butanestannonic acid	4000	r
Octyltin trichloride	2400–3800	r
Octyltin tri(isooctylthioglycolate)	3400–4000	r

[a] A range indicates highest and lowest values for LD_{50} reported in the literature

gical data for organotins has been compiled by Smith [200]. Table 8 lists acute oral LD_{50} values for some organotin compounds.

Progressive introduction of organic groups at the tin atom in any R_nSnX_{4-n} series produces a maximum biological activity against all species when n is 3, *ie* for the triorganotins, R_3SnX. If the chain length of the n-alkyl group is increased within any trialkyltin series, highest mammalian toxicity is obtained for the triethyltin compounds. As the chain length increases beyond butyl, a sharp drop in biological activity occurs and the tri-n-octyltin compounds are essentially nontoxic to all living organisms [200]. The species towards which the R_3SnX compound is most active is primarily determined by the nature of the three organic groups R (Table 9) An effect of particular environmental importance is the high toxicity of tributyl-, triphenyl- and tricyclohexyl-tin compounds to fish and other marine creatures, and in this context a bibliography of the toxicity of organotin compounds to aquatic animals has been compiled [211].

As with the trialkyltins, the mammalian toxicity of dialkyltin compounds, R_2SnX_2, generally decreases with increasing length of the alkyl chain, and the mono-organo-

Table 9. Species dependence on nature of R group in R_3SnX [201]

R.	species
methyl	Insects
ethyl	Mammals
n-propyl	Gram-negative bacteria
n-butyl	Gram-positive bacteria, fish, fungi, molluscs
phenyl	Fish, fungi, molluscs
cyclohexyl } neophyl }	Mites

tins, $RSnX_3$, which do not appear to have any important toxic action in mammals, show the same pattern.

Animal Studies

Toxicity of tetra-alkyltins to animals is characterised by a slow development of symptoms and the lengthy interval which often elapses between administration and animal mortality. Symptoms of poisoning are similar to those for the trialkyltin analogue [224]. It seems probable that tetra-alkyltin compounds are converted to trialkyltins in the liver and this has been demonstrated experimentally by Cremer [86], who detected triethyltin in the tissues of rats and rabbits injected with tetra-ethyltin.

Stoner et al. [207] first demonstrated the toxicity of trialkyltin compounds in warm-blooded animals in 1955. In oral administration to rats, of the compounds tested, triethyltin sulphate had the lowest LD_{100} (10 mg/kg body weight); the corresponding dose for trimethyltin sulphate was 20 mg/kg. As the chain length of the alkyl groups increased beyond ethyl, toxicity decreased; thus trihexyltin acetate had an LD_{100} in excess of 100 mg/kg. Rats poisoned with short chain length trialkyltin compounds initially show a characteristic muscular weakness. After a period of apparent recovery, weakness reappears and continues until all the rats die. In rabbits convulsions occur after the initial recovery period. The principal site of attack is the central nervous system and severe oedema of the white matter of the brain is the major pathological finding. In both animals and man, mild cases of poisoning recover completely.

The effect of triethyltin compounds (20 ppm in the diet) on the water content, electrolyte distribution and kinetics of radioactive sodium, chloride and sucrose uptake of cerebral cortex and skeletal muscle has been studied by Reed et al. [179]. Water content of the cortex was increased by 1.2%, but muscle water was not altered. Potassium levels were not changed in brain cortex and skeletal muscle; sodium and chloride increased in both of these tissues. Triethyltin-induced cerebral oedema in rats, characterised by a reduction in extracellular space and increase in glial cell volume, is apparently due to an increased permeability of the glial cell membrane to sodium and chloride, resulting in an influx of both ions. A toxic effect observed for trimethyltin compounds is that they cause selective and irreversible neuronal destruction in the brain [7, 57, 184].

The lower trialkyltin compounds are not used commercially because of their toxicity. However, tributyltin compounds such as bis(tributyltin) oxide are used

Table 10. Findings in feeding studies with bis(tributyltin) oxide in juvenile rats [187]

Study period	Animals/ group	Dose (ppm)	Salient findings
4 weeks (dose range-finding)	5 males and 5 females	4	no substance-related effects observed
		20	no substance-related effects observed
		100	food consumption and weight gain, absolute thymus weight (M)
		500	high mortality, apathy, emaciation, thymus and lymph node weight (M + F), depletion of lymphocytes in lymphatic organs, atrophy of thymus and lymph node
13–14 weeks (subchronic study)	10 males and 10 females	0	no effects observed
		4	no substance-related effects observed
		20	slight prolongation of coagulation times (M), food consumption $\downarrow$ (F)
		100	food consumption and weight gain $\downarrow$, serum alkaline phosphatase $\uparrow$ albumin $\uparrow$ (F), γ-globulin $\downarrow$ (F), weight of thymus, lymph node and thyroid $\downarrow$, adrenal weight $\downarrow$ (M)

M: male animals, F: female animals
$\uparrow$: increase, $\downarrow$: decrease
ppm: mg/kg diet

extensively as industrial biocides, for example in wood preservation and disinfection, and the toxicity of tributyltin compounds has been reviewed in some detail by Schweinfurth [187]. Vapours of bis(tributyltin) oxide, tributyltin naphenate and tributyltin benzoate proved to be practically non-toxic in single exposure studies with rats. After repeated administration of bis(tributyltin) oxide to rats in the feed (90 days) and in aerosol form (28 days), toxicity to lymphatic organs predominates (Table 10). After aerosol exposure there is also severe irritation of the respiratory tract. There was no indication of damage to the central nervous system akin to that which occurs with the lower trialkyltin homologues. In a 4 week inhalation study, no harmful reactions were observed in rats exposed to vapour practically saturated with bis(tributyltin) oxide.

Symptoms of poisoning produced by dialkytin compounds are completely different from those caused by trialkyltins. The main effect is on the liver and bile duct. Severe lesions are produced in the bile duct by a toxic dosage and death results from hepatic damage or peritonitis [30]. The most pronounced effect occurred with dibutyltin compounds; dimethyltin compounds did not produce this effect [29]. Dioctyltin compounds have generally low toxicities and certain compounds (mercaptides and carboxylates) have been approved for use as stabilisers for PVC for food applications.

Although mono-alkyltin compounds have very low toxicities, Pelikan and Cerny [171, 172] administered large (4000 mg/kg) single doses of monobutyltin trichloride, monobutyltrin tris (2-ethylhexyl mercaptoacetate), monobutyltin triacetate and monon-octyltin tris(2-ethylhexylmercaptoacetate) to male and female mice. After 24 hours, the mice did not respond to light, sound or mechanical stimuli. All mice were killed 24 hours after testing began and the organs examined. Fatty degeneration of the liver and kidneys was reported for all compounds.

Acute LD_{50} values for triphenyltin compounds fed orally to rats are in the range 125–150 mg/kg (ie similar in toxicity to bis(tributyltin) oxide which has LD_{50} values of 150–200 mg/kg [170]). Toxic effects of triphenyltins after oral administration appear only slowly; they include weakness, anorexia, rough coat, watery diarrhoea, staggering and death in coma [135]. The main action is thought to be on the central nervous system, but in contrast to triethyltin compounds, cerebral oedema does not occur [63]. The dermal LD_{50} of triphenyltin acetate for rats was found to be about 450 mg/kg body weight [135]. Chronic toxicities have been determined by short- and long-term feeding studies with triphenyltin compounds on rats, guinea pigs and dogs [48]. No-effect levels of 1.0–5.0 ppm were derived for rats, 5.0 ppm for guinea pigs and 2.0 ppm for dogs. In reproduction experiments, Scholz and Baeder [185] fed concentrations of 25 ppm triphenyltin acetate to SPF-Wistar-K rats; there were no adverse effects on general health, fertility, or postnatal developments of rats in the generations examined. At higher doses (250 ppm) some adverse effects were noted. Trials with rats fed 1,2 or 5 ppm of triphenyltin hydroxide indicated no negative influence on fertility, number of young in litter, vitality, body weight or mortality of offspring in 3 consecutive generations [15].

Effects in Humans

There are relatively few clinical observations and epidemiological data on the effects of organotin compounds in man, which is perhaps indicative of their safety when used under stated conditions and with adequate precautions, such as protective clothing. There have been, however, a few cases of occupational exposure, including 2 deaths, and these instances are summarised in Table 10.

Di- and tri-butyltin compounds have caused acute skin burns on laboratory and process workers and these appear 1–8 hours after exposure [145]; when the substance is washed off immediately no lesion occurs. A more diffuse, but less rapidly healing erythematous eruption occurs through contact with clothes moistened with the vapour or liquid of organotin compounds. Following contact with the eye, lachryma-tion and severe suffusion of the conjunctivae occurred within a few minutes and persisted for 4 days.

As shown in Table 11, degrees of intoxication ranging from slight to serious and even to mortality, have been reported as a result of occupational exposure. In most of these cases, implementation of specified safety procedures or the wearing of suit-able clothing, would have obviated the risk of incident.

The only epidemiological incident which dates back to the early use of organotins, illustrates the need for careful manufacturing controls and thorough testing of newly introduced formulations. In the incident which occurred in France in 1954, oral administration of a proprietary formulation for boils, based on diethyltin di-iodide and linoleic acid, led to at least 100 deaths and over 200 intoxications. The clinical data have been reviewed by Alajouanine et al. [4]. It is probable that most of the symptoms could be attributed to a cerebral oedema of the white matter of the

Table 11. Accidents reported involving organotin compounds

Compound	Number and sex of workers	Description of exposure	Effects	Ref.
Triphenyltin acetate	1M	Spraying sugar beets with aqueous triphenyltin acetate solution for 2 hours	Violent headaches, unconsciousness	109
Triphenyltin acetate	2M	Formulating fungicidal spray solution based on triphenyltin acetate	Vomiting, shortness of breath, glyosuria in one case; violent headache, nausea, vomiting, epigastric pain in the other	109
Triphenyltin acetate	48M	Weighing and bagging a commercial pesticide based on triphenyltin acetate (8 hours/day for 2—10 days)	Irritation of skin, mucous membranes, conjunctivae	154
Triphenyltin acetate	2M	Aerial spraying of triphenyltin acetate based pesticide	Dyspepsia, severe diarrhoea, blurred vision, liver damage in one case; heartburn, blurred vision, diarrhoea, coughing and hyperglycaemia in the other	
Triphenyltin acetate	1M	Loading plane with triphenyltin acetate based pesticide	Skin irritation, dizziness, headache, nausea, fatigue, chronic hepatitis	159
Triphenyltin chloride	1F	Drenched with hot slurry containing triphenyltin chloride whilst working in organotin manufacturing plant	Severe burns, death	19
Bis(tributyltin)oxide	45 —	Construction of sonar domes using rubber containing bis(tributyltin)oxide. Air concentrations 0.1—0 mg/m^3 (as tin metal)	Irritation of eyes and upper respiratory tract	19
Bis(tributyltin)oxide	—F	Spray painting with latex paint containing bis(tributyltin)oxide	Irritation of eyes and nasal mucosa	139
Tributyltin chloride and dibutyltin dichloride	—	Employed in organotin manufacturing plant	Dermatitis	
Trimethyltin chloride and dimethyltin dichloride	2M	Vapours released into workplace during operation of pilot plant for synthesis of dimethyltin dichloride from metallic tin and monochloromethane	Mental confusion, with generalised epileptic seizures. Complete recovery following removal from exposure	103
Trimethyltin chloride and dimethyltin dichloride	6M	Cleaning a cauldron containing tri-methyltin and dimethyltin chlorides. Exposed to vapour and liquid whilst allegedly wearing protective equipment (Max 9 × 10 minutes over 3 days)	Cerebral oedema with initial symptoms of headaches, tinnitis, deafness, disorientation and in most severe cases, respiratory depression requiring ventilator assistance. 1 mortality, 2 still under treatment after 3 years, 3 clinically healthy	180

brain. The symptoms and clinical findings seemed to favour a trialkyltin compound as causative agent, probably acting synergistically with other constituents. It appears likely that toxic triethyltin iodide was present in the formulation as an impurity. Since that time, more rigorous controls have been imposed on the manufacture and use of organotin compounds and in over 30 years of use, no other incidents of this kind have been reported.

Mode of Toxicological Action

The underyling cause of the broad spectrum of biocidal activity exhibited by the triorganotin compounds is thought to be derangement of mitochondrial functions [56, 191]. The mechanisms involved are:

a) Interaction with mitochondrial membranes to cause swelling and disruption.
b) Secondary effects due to their ability as ionophores to derange mitochondrial function through mediation of Cl^-/OH^- exchange across the lipid membrane.
c) Inhibition of the fundamental energy conservation process involved in the synthesis of ATP from ADP, for example by mitochondrial oxidative phosphorylation and also photosynthetic phosphorylation in chloroplasts.

Stockdale et al. [206] showed that the order of effectivness of organotin compounds in inhibiting coupled respiration was tributyl $>$ tripropyl $>$ triphenyl $>$ trimethyl. Two separate effects were suggested (a) an oligomycin-like inhibition of coupled phosphorylation and (b) an alteration of hydroxide exchange across lipid membranes, producing uncoupling, swelling and reduction of intramitochondrial substrate and phosphate concentrations, followed by structural damage.

On a macromolecular scale, a number of researchers have attempted to identify possible binding sites for triorganotins. Trialkyltin compounds have been shown to bind to certain proteins and the exact chemical nature of the binding sites has been much studied [5, 64, 102, 191]. Aldridge found [5] that two males of triethyltin chloride are bound to one mole of either cat or rat haemoglobin and that at least one of these binding sites involves histidine. Cardarelli [68] conducted in vitro studies with bis(tributyltin) oxide in snail protein and found a high order of activity with amino acids; he suggested that mortality arose from direct reaction between organotins and proteins. Rose and Lock [181, 183] proposed a structural model based on a planar R_3Sn moiety bridged axially by two histidine molecules, but more recent [119m]Sn Mossbauer investigations of triethyltins bound to high affinity protein sites indicate either a tetrahedral R_3SnX tin atom geometry or a pentacoordinate cis-R_3SnX_2 chelated structure [94, 102].

Much less is known about other possible binding sites for triorganotins on protein, although work by Limouzin [142], Griffiths [64] and Tan [213] and their co-workers suggests sulphydryl groups as possible points of attachment and this was confirmed in a study [212] of the binding of triethyltin bromide to cat haemoglobin, which showed that cysteine as well as histidine residues were involved:

Like the trialkyltin compounds, triphenyltins inhibit oxidative phosphorylation of ADP to ATP in mitochondria [192, 206] and chloroplasts. Inhibition of the Na^+-K^+ ATP-ase in cell membranes and of ATP-ase for Ca^{2+} transfer in sarcoplasmic reticulum has been found [191].

An intramolecularly chelated tributyltin dialkylamino-alkoxide has been found to be much less toxic to mice than bis(tributyltin) oxide and it is considered [223] that the toxicity of the trialkyltin compounds R_3SnX may be independent of the X radical only when this is a simple non-chelating group which is capable of exchanging at these active protein sites. Further evidence for this was obtained by Blunden et al. [47] who found that with X groups which result in the formation of an intramolecularly chelated five coordinate tin atom geometry

a significant reduction in biological activity may occur.

The biological effects of the diorganotin compounds R_2SnX_2 are also due to their ability to inhibit oxygen uptake in mitochondria. However, the toxic mechanism is different since the diorganotins have been shown to inhibit α-ketoacid oxidation [5], probably by combining with enzymes or co-enzymes possessing vicinal dithiol groups, eg reduced lipoic acid:

This means that the nature of the X groups can again affect the toxicity, for example, dimethyltin dichloride is moderately toxic (since it can react readily with dithiol moieties) whereas dimethyltin diiso-octyl thiogycolate is relatively non toxic [200] (as it already possesses two Sn-S bonds in the molecule). One particular biological effect of certain dialkyltins, noted by Seinen [189, 190] is their toxicity towards lymphocytes and the suppression of thymus dependent immunity in rats. This appears to be more marked for compounds of longer chain length and is in contrast to the general pattern of acute toxicities as expressed by their LD_{50} values [200]. In a more recent paper Penninks and Seinen [172] have reviewed the immuno-toxic effects of dialkyltin compounds in rats. They induce lymphocyte depletion in the thymus and the thymus dependent areas of the peripheral lymphoid organs without there being any signs of a generalised toxicity. Because of their selective lymphocytotoxic action, they cause immuno-suppression especially of the cell-mediated immunity. The immunotoxicity is suggested as being due to the affinity of dialkyltin compounds for dithiol groups in plasma membrane and/or cytoskeletal proteins, disturbing cell energetics as well as cell proliferation.

Metabolism of Organotins

From an environmental standpoint, an understanding of the metabolic pathways for organotin compounds in mammals is of great importance. Cremer [87] was the first to demonstrate that tetraethyltin is metabolised in vivo in rats to a triethyltin derivative and it has also been shown that diethyltin species are broken down in vivo in rats to monoethyltin salts, which are rapidly eliminated from the body [52]. Wada et al. studied [28, 124, 126, 230] the in vivo metabolism in rabbits of a series of tetraalkyltin compounds and found that the extent of conversion of R_4Sn to R_3SnX decreased as the length of the alkyl chain increased [28].

In vivo studies in rats and mice, using tricyclohexyltin [133], triethyltin [126] and tributyltin [11, 59, 100, 126] compounds have indicated that they are excreted essentially intact from the animal so that there seems to be no problem of long-term accumulation. However, Blair [36] reported that tricyclohexyltin hydroxide is metabolised in vivo in rats, to di- and monoorganotin and inorganic tin species and Iwai et al. [125]

Table 12. Concentration of organotins in rat urine after dosing with triphenyltin compounds [105]

Compound	Radioactivity % of total
$(C_6H_5)_3Sn^+$	25
$(C_6H_5)_2Sn^{2+}$	17—19
$(C_6H_5)Sn^{3+}$	20—29
Inorganic tin (IV)	29—33
Not extracted from aqueous solution	0.6—5

demonstrated that in rats, tributyltin fluoride, once transported to the liver, is dealkylated. Freitag and Bock [105] examined the metabolism of triphenyltin chloride in rats by dosing them with a single oral dose of 3 mg of a tin-113 labelled compound and analysing degradation products (Table 12).

Fish and his co-workers [134] also studied the breakdown of tin-113 labelled triphenyltin acetate in rats and found a similar breakdown pattern of the triaryltin compound. Rose and Aldridge [182] examined distribution of triethyltin in various tissues of rats, guinea pigs and hamsters, after injection of tin-113 triethyltin (10 mg/kg) (Table 13).

It is interesting to see variations in the distribution pattern for different animals; for the rat, highest tin concentrations appear in the blood, whereas for hamster and guinea pig, highest levels are in the liver whilst the bile in these cases contained copious amounts of tin. This finding is ascribed to the efficient binding of tin to rat haemoglobin.

In vitro animal studies have demonstrated, using rat liver microsomal monooxygenase enzyme systems, that the primary metabolic reaction for butyltin compounds is not Sn-C bond cleavage, but carbon hydroxylation of the butyl groups, where the α and β carbon-hydrogen bonds are found to be more susceptible to hydroxylation [134]. Similarly, results indicating hydroxylation of the organic group have been obtained with tricyclohexyltin hydroxide [133], but this process does not appear to occur with triphenyltin species [134].

In the majority of studies carried out on the metabolism of organotin compounds, little information has been derived regarding the dealkylation or dearylation products formed from the organic moieties. However, work by Prough et al. [177] on the NADPH- and oxygen-dependent microsomal metabolism of the ethyltin series Et_nSnX_{4-n} (n is 2–4) has demonstrated that the major organic metabolite is ethylene and the minor product ethane. Similarly, Fish and his co-workers [134] detected 1-butene from the butyltin compounds, Bu_4Sn and Bu_3SnX (where X is chloride, acetate and $OSnBu_3$).

Table 13. Levels of triethyltin in animal tissues four hours after injection of tin-113-triethyltin (10 mg/kg) (animals perfused before tissue samples collected) [182]

Tissue	Triethyltin (Nanomoles/g Wet Weight)		
	Rat	Guinea pig	Hamster
Liver	103.0	275.0	124.0
Kidney	52.7	78.5	31.4
Muscle	16.2	17.8	26.0
Brain			
Cerebellum	22.4 }	28.1	} 25.2
Brain stem	22.0 }		}
Cerebrum (half)	22.4	—	—
Cortex	21.2	—	—
Remainder of brain	21.6	24.8	24.6
Spinal cord	20.7	24.4	23.1
Fat	30.4	11.2	48.7
Bile	—	57.0	404.0
Blood	224.0	8.3	7.7

Carcinogenic Potential

It is important to note that no carcinogenic effects have yet been demonstrated for any of the organotin compounds tested to date (Table 14). Studies on carcinogenicity of organotin compounds are in progress at the Rijksinstitut voor de Volksgezondheid in the Netherlands and a corresponding study on mice began in 1985: Schweinfurth [187] has reported that results of testing for genotoxic potential in micro-organisms and in the micronucleus test in mice, did not indicate any mutagenic or carcinogenic potential.

Table 14. Carcinogenicity studies on organotin chemicals

Compound	Species	Type of test	Result	Ref.
Ph_3SnOAc	Mice (Male and female)	18 month feeding study	Negative	123
Ph_3SnOH	Mice (Male and female)	78 week feeding study	Negative	22
Ph_3SnOH	Rats (Male and female)	78 week feeding study	Negative	22
Bu_3SnF	Mice (Male)	6 months dermal study	Negative	194
$(cyclo\text{-}C_6H_{11})_3SnOH$	Rats (Male and female)	2 year feeding study	Negative	14
$Bu_2Sn(OAc)_2$	Mice (Male and female)	78 week feeding study	No conclusive evidence for carcinogenicity	23
$Bu_2Sn(OAc)_2$	Rats (Male)	78 week feeding study	No conclusive evidence for carcinogenicity	23
$[(PhMe_2CCH_2)_3Sn]_2O$	Mice (Male and female)	18 month feeding study	Negative	20

Preventive Measures

The number of industrially used organotin compounds with high toxicity is surprisingly low. Nevertheless, certain precautions are needed in the manufacture, handling and disposal of organotins. Advice on toxicology and safe handling of organotin compounds is often published by manufacturers, for example, the booklet by Schering AG [12]. The US Department of Health, Education and Welfare's National Institute of Occupation Safety and Health (NIOSH) has published criteria for a recommended standard on occupational exposure to organotin compounds [19]. This recommends control of organotin concentrations in the workplace so that they remain at or below 0.1 mg/m^3 air and facilities for sampling and analysis should be present. Operatives who may be occupationally exposed to organotins should be provided with regular medical surveillance. In areas where organotins are being used, a sign containing information on effects of specific compounds on human health should be prominently displayed; containers should be adequately labelled and kept tightly closed when not in use. Solids are obtained as powder or pellets and are best transferred to containers under a dust collector. Liquid organotin compounds are best transferred under nitrogen. Inhalation of vapour must be prevented by suitable exhausts or by use of respirators. Stainless steel has proved a suitable material for tanks, pipelines and apparatus for containing organotin compounds; however, stainless steel does not resist alkyltin halides above 150 °C. All liquid organotin compounds may also be kept in steel tanks coated with synthetic resin-based paints. For exposure to elevated temperatures, glass and vitreous enamel have proved the most suitable material for all organotins.

Protective clothing, including face shields, goggles, protective boots, etc should be provided for operatives and used according to organotin suppliers' instructions. Respirators should be worn when there is inhalation risk, for example during spraying antifouling paint or preservative. Before maintenance begins in organotin plants, sources of the compounds should be isolated. Employees who have had skin contact with organotins should immediately wash and shower for at least 15 minutes to remove all traces from the skin; contaminated clothing should be removed and cleaned, or disposed of. Eating, drinking and smoking should not be permitted in areas where organotins are used, transferred, stored or manufactured.

Disposal of Waste

Disposal of organotin wastes has been discussed by Plum [175] and must be in accord with local or government regulations. If large amounts of organotin compounds reach water, removal should be attempted as far as possible, by skimming or draining off; the remaining or dissolved parts can be removed by chemical means or through absorption. Organotin compounds can be decomposed to inorganic forms by oxidation, using potassium permanganate in acid solution at room temperature. Certain activated carbons have proved suitable for removing organotin compounds from polluted water. Large amounts of organotin waste and adsorption material containing organotins can be destroyed by burning in special plants or taken to special refuse tips. If large amounts of liquid organotin compounds have entered the soil, this soil must be removed and either incinerated or taken to special refuse tips. Pollution of underground waters must be avoided at all costs. Spilled organotin compounds, eg bis(tributyltin) oxide, can be adsorbed with sawdust or fine sand which is then disposed of in the manner described.

Future Trends

Usage of organotin compounds has grown over the last 30 years from about 1500 to 35,000 tonnes a year and these compounds are now employed in a wide range of industries. A tremendous body of published research work on these compounds underlies their use and this includes extensive data on their toxicology. Careful adherence to recommended practices has meant that over this period, no serious disasters involving organotins have been reported. When looking at the environmental future of these compounds, a number of factors have to be borne in mind.

The first is the sheer increasing tonnage of compounds which are being introduced into the environment. An example exists in the field of antifouling coatings, where the proliferation of small pleasure boats protected with paints containing organotins has already resulted in significantly increased levels of organotins in harbours and marinas (see page 5). Deleterious effects have been observed in Europe on certain species of oyster and legislation is currently being introduced to limit this influence. The development and increasing use of polymer-based antifouling paints is likely to keep this release of organotin toxicant within acceptable and manageable proportions.

A second factor to be considered is the development of new organotin compounds with industrial potential. Whereas the toxicity of conventional mono-, di-, tri- and tetra-organotins is well documented and the effect of types of alkyl or aryl group and anionic constituent is understood, more complex variations on these compounds will need to be evaluated, and also possible synergism between active constituents in combined formulations. There is a safety net here, since before a new compound receives approval for commercial use, it has to undergo an extremely long period of testing. Huber [121] considers that 8–10 years is the likely timescale for the first kilogramme of new compound to be ready for commercial trials in agricultural applications.

Another important consideration is the rapid development of analytical techniques for determining extremely low concentrations of organotin compounds in environmental samples. This has led to traces of organotins, for example, methyltins, being reported in lakes and rivers in many different regions. Concern over build-up of organotins (and other heavy metals) in the environment will undoubtedly lead to improved methods of use for these compounds, to make them effective at lower levels. One approach is the slow-release concept whereby toxicants are encapsulated in matrices from which they are released at low, controlled rates. Others include the use of triorganotins as antifeedants for pest control and in integrated control procedures alongside the use of natural predators to protect crops.

Concern over mutagenic or carcinogenic behaviour of organotins will hopefully be allayed when the results of long-term tests now in progress become available. Certainly, results obtained to date have not indicated such effects in commercially used organotins.

In 1978 the world's manufacturers set up a body, the Organotin Environmental Programme (ORTEP) Association, whose aim is to promote and foster the dissemination of scientific and technical information on the environmental effects of organotin compounds. Members of the Association meet regularly and a Data Bank on environmental aspects of organotins is maintained. The existence of bodies such as the ORTEP Association will also ensure that information on environmental behaviour of organotin compounds will be made freely available as it accrues. Whereas in the past many chemical agents, proven effective on empirical grounds, have only later been shown to have harmful effects, in the case of organotin compounds, their use has from the start, been carefully planned and controlled and supported by extensive research.

References

1. Adema, C. M.: David Taylor Naval Ship Research and Development Center, Rep. 79—041, Mar (1979)
2. Adema, C. M., Schatzburg, P.: Naval Engrs. J., May, 209 (1984)
3. Akagi, H., Sakagami, Y.: Bull. Inst. Publ. Health Jpn., *20*, 1 (1971)
4. Alajouanine, T., Derobert, L., Thiefry, S.: Rev. Neurol., *98*, 85 (1958)
5. Aldrige, W. N.: in 'Organotin Compounds: New Chemistry and Applications', ed. Zuckerman, J. J., Am. Chem. Soc. Adv. Chem. Ser., *157*, 186 (1976)
6. Aldridge, W. N.: in 'Proc. Int. Conf. Si, Ge, Sn and Pb Compounds', eds. Gielen, M., Harrison, P. G., Rev. Si, Ge, Sn and Pb Compds, Special Issue, Freund Publishing House, Tel Aviv, 9 (1978)
7. Aldridge, W. N., Brown, A. W., Brierley, J. B., Verschoyle, R. D., Street, B. W.: Lancet, 692 (1981)
8. Aldridge, W. N., Street, B. W.: Analyst, *106*, 60 (1981)
9. Allen, D. W., Brooks, J. S., Unwin, J., McGuinness, J. D.: Chem. Ind., 5th Aug, 524 (1985)
10. Alzieu, C., Thibaud, Y., Heral, M., Boutier, B.: Rev. Trav. Inst. Pêches Marit., *44*, 301 (1980)
11. Anger, J. P.: Ph. D. Theis, Univ. René Descartes de Paris (1975)
12. Anon: 'Advice on Toxicology and Safe Handling of Organotin Compounds', Publn, Schering AG. Bergkamen, 20 pp
13. Anon: M and T Chemicals Inc., Standard Testing Method, AA-33
14. Anon: '1970 Evaluations of Some Pesticide Residues in Food', FAO/WHO, Rome, 521 (1971)
15. Anon: FAO/WHO, Rome 1971, AEP: 1970/M/12/1, 327 (1971)
16. Anon: 'Vendex Miticide', Tech. Data Bull., Shell Chemicals Co., San Ramon, CA (1973)
17. Anon: '1973 Evaluations of Some Pesticide Residues in Food', WHO Pesticide Residue Ser., No. 3, 448 (1974)
18. Anon: WHO Tech. Rep. Ser. No 545, 440 (1974)
19. Anon: U.S. Dept. Health Educat. Welfare, Nat. Inst. Occup. Safety Health, Nov. 187 pp (1976)
20. Anon: '1977 Evaluations of Some Pesticide Residues in Food', FAO/WHO, Rome, 229 (1977)
21. Anon: 'Tributyltin Oxide, Notes on Handling and Toxicity', Tech. Service Note, Albright and Wilson, Oldbury (1979)
22. Anon: U.S. Natl. Cancer Inst., Carcinogen Tech. Rep. Ser., No 139 (1979)
23. Anon: U.S. Natl. Cancer Inst., Carcinogen Tech. Rep. Ser., No 183 (1979)
24. Anon: 'Specifications and Technical Data, Organotin Compounds', Schering AG, Bergkamen (1981)
25. Anon: U.K. Dept. Environ., Press Notice 373, 24th July (1985)
26. Anon: Water Bull., 23rd Aug., 13 (1985)
27. Arakawa, Y., Wada. O., Manabe, M.: Anal. Chem., *55*, 1901 (1983)
28. Arakawa, Y., Wada. O., Yu. T. H.: Toxicol. Appl. Pharmacol., *60*, 1 (1981)
29. Barnes, J. M., Magee, P. N.: J. Pathol. Bacteriol., *75*, 267 (1958)
30. Barnes, J. M., Stoner, H. B.: Brit. J. Ind. Med., *15*, 15 (1958)
31. Barnes, R. D., Bull. A. T., Poller, R. C.: Pestic. Sci., *4*, 305 (1973)
32. Barug, D.: Chemosphere, *10*, 1145 (1981)
33. Barug, D., Vonk, J. W.: Pestic. Sci., *11*, 77 (1980)
34. Beiter, C., Engelhart, J. E., Freiman, A., Sheldon, A. W.: Proc. Am. Chem. Soc., Symp. Mar. Freshwater Pestic., Atlantic City, NJ, Aug (1974)
35. Bennett, R. F.: R.S.C. Ind. Bull., *2*, 171 (1983)
36. Blair, E. H.: Environ. Qual. Safety Suppl., *3*, 406 (1975)
37. Blair, W. P., Olsen, G. J., Brinckman, F. E., Iverson, W. P.: Microb. Ecol., *8*, 241 (1982)
38. Blair, W. P., Jackson, J. A. Olsen, G. J., Brinckman, F. E., Iverson, W. P.: 3rd Internat. Conf. Heavy Metals in the Environment, 235 (1981)
39. Blunden, S. J.: J. Organometal. Chem., *248*, 149 (1983)
40. Blunden, S. J., Chapman, A. H.: Analyst, *103*, 1266 (1978)
41. Blunden, S. J., Chapman, A. H.: 'Review of the Degradation of Organotin Compounds in the Environment and Their Determination in Water', Int. Tin Res. Inst., Publ. No 623, 7 pp (1983)

42. Blunden, S. J., Chapman, A. H.: in 'Organometallics in the Environment' ed. Craig P. J., Longma Group, London, Vol 3 (1986)
43. Blunden, S. J., Cusack, P., Hill, R.: 'The Industrial Uses of Tin Chemicals', Royal Society of Chemistry, London (1985)
44. Blunden, S. J., Hill, R.: J. Organometal. Chem., 267, C5 (1984)
45. Blunden, S. J., Hill, R.: Inorg. Chim. Acta, 87, 83 (1984)
46. Blunden, S. J., Smith, P. J., Gillies, D. G.: Inorg. Chim. Acta, 60, 105 (1982)
47. Blunden, S. J., Smith, P. J., Sugavanam, B.: Pestic. Sci., 15, 253 (1984)
48. Bock, R.: 'Residue Reviews', ed. Gunther, F. A., Springer-Verlag, New York, Vol 79 (1981)
49. Bokranz, A., Plum, H.: Topics Current Chem., 16, 365 (1971)
50. Bollen, W. B., Tu, C. M.: Tin Its Uses, 94, 13 (1972)
51. Braman, R. S., Tompkins, M. A.: Anal. Chem., 51, 12 (1979)
52. Bridge, J. W., Davis, D. S., Williams, R. T.: Biochem. J., 105, 1261 (1967)
53. Brinckman, F. E., Iverson, W. P.: 169 Am. Chem. Soc. Meeting, Philadelphia, PA, 8–10 Apr (1975)
54. Brinckman, F. E., Olsen, G. J., Thayer, J. S.: paper presented to the Symposium on Marine Estuarine Geochemistry, Internat. Chem. Congress of Pacific Basin Societies, Honolulu, Hawaii, Dec. 16—21 (1984)
55. Brinckman, F. E., Parris, G. E., Blair, W. R., Jewett, K. L., Iverson, W. P., Bellama, J. M.: Environ. Health Perspect., 19, 11 (1977)
56. Brooks, J. S., Allen, D. W., Unwin, J.: Polym. Degrad. Stab., 10, 79 (1985)
57. Brown, A. W., Aldridge, W. N., Street, B. W.: Neuropath. Appl. Neurobiol., 5, 83 (1979)
58. Brown, R. A., Nazario, C. M., de Tirado, R. S., Castrilton, J., Agard, E. T.: Environ. Res., 13, 56 (1977)
59. Brüggemann, J., Klimmer, O. R., Niesar, K. H.: Zentralbl. Veterinar. Med., 11, 40 (1964)
60. Brüggemann, J., Barth, K., Niesar, K. H.: Zentralbl. Veterinar. Med., 11, 4 (1964)
61. Bunyatyan, Y. A., Oganesyan, G. O.: Gig. Sanit., 3, 55 (1983)
62. Byrd, J. T., Andreae, M. O.: Science, 218, 565 (1982)
63. Cahen, R., Boucard, M., Lalouric, M., Lacour, C.: Compt. Rend. Acad. Sci., 271, Ser. D. 1816 (1970)
64. Cain, K., Griffiths, D. E.: Biochem. J., 162, 575 (1977)
65. Cain, K., Pastis, M. D., Griffiths, D. E.: Biochem. J., 166, 593 (1977)
66. Cardarelli, N. F.: in 'Molluscicides in Schistosomiasis Control', ed. Cheng, T. C., Academic Press, New York, 177 (1974)
67. Cardarelli, N. F.: 'Controlled Release Pesticide Formulations', Chemical Rubber Co., Cleveland, OH (1976)
68. Cardarelli, N. F.: 'Controlled Release Molluscicides', Environ. Manage. Lab. Mon., Akron, OH, 34 (1977)
69. Carey, J. K.: Ph. D. Thesis, Univ. London (1980)
70. Carr, H. G.: Soc. Plast. Engrs., 25, 72 (1969)
71. Cenci, P., Cremonini, B.: Ind. Sacc. Ital., 62, 313 (1969)
72. Challenger, F.: in 'Organometals and Organometalloids, Occurence and Fate in the Environment', eds. Brinckman, F. E., Bellama, J. M., Am. Chem. Soc. Symp. Ser., 82, 1 (1978)
73. Chapman, A. H.: Unpublished Work, Int. Tin. Res. Inst. Perivale, Middx (1981)
74. Chapman, A. H., Price, J. W.: Int. Pest Control, 14 (Jan/Feb), 11 (1972)
75. Chapman, A. H., Price, J. W.: Unpublished Work, Int. Tin. Res. Inst., Perivale, Middx (1973)
76. Chau, Y. K., Wong, P. T. S., Bengert, G. A.: Anal. Chem., 54, 246 (1982)
77. Chau, Y. K., Wong, P. T. S., Krama, D., Bengert, G. A.: in 'Proc. Int. Conf. Heavy Metals in the Environment (Amsterdam)', ed. Ernst, W. H., CEP Consultants, Edinburgh (1981)
78. Chromy, L., Uhacz, K.: J. Oil Col. Chem. Assoc., 51, 494 (1968)
79. Chromy, L., Uhacz, K.: J. Oil Col. Chem. Assoc., 61, 39 (1978)
80. Coussement, S.: Ann. Gembloux, 78, 41 (1972)
81. Craig, P. J., Rapsomanikis, S.: in 'Proc. DOE/NBS Workshop Environ. Spec. Monitor Needs', eds. Brinckman, F. E., Fish, R. H., NBS Spec. Publ. No 618, U.S. Dept. Commerce, Washington, DC., 54 (1981)
82. Craig, P. J., Rapsomanikis, S.: J. Chem. Soc., Chem. Commun., 114 (1982)
83. Craig, P. J., Rapsomanikis, S.: Environ. Tech. Letts., 5, 407 (1984)

84. Craig, P. J., Rapsomanikis, S.: Inorg. Chim. Acta, *107*, 39 (1985)
85. Craig, P. J., Rapsomanikis, S.: Environ. Sci. Technol., *19*, 726 (1985)
86. Cremer, J. E.: Biochem. J., *67*, 28 (1957)
87. Cremer, J. E.: Biochem. J., *68*, 685 (1958)
88. Davies, A. G., Goddard, J. P., Hursthouse, M. B., Walker, N. P. C.: J. Chem. Soc. Chem. Commun., 597 (1983)
89. Davies, A. G., Smith, P. J.: in 'Comprehensive Organometallic Chemistry', eds. Wilkinson, G., Stone, F. G. A., Abel, E. W., Pergamon Press, Oxford, *2*, 519 (1982)
90. Dizikes, L. J., Ridley, W. P., Wood, J. M.: J. Am. Chem. Soc., *100*, 1010 (1978)
91. Dowd, T.: U.S. Patent 4 261 187 (1980)
92. Dudley-Brendall, T. E., Dickinson, D. J.: Int. Res. Group Wood Pres., Doc. No IRG/WP/1156 (1982)
93. Duncan, J.: Pharmacol. Ther., *10*, 407 (1980)
94. Elliot, B. M., Aldridge, W. N., Bridges, J. W.: Biochem. J., *177*, 461 (1979)
95. Evans, C. J.: Tin Its Uses, *86*, 7 (1970)
96. Evans, C. J.: Tin Its Uses, *110*, 6 (1976)
97. Evans, C. J., Hill, R.: Rev. Si, Ge, Sn and Pb Compds., *7*, 57 (1983)
98. Evans, C. J., Karpel, S.: 'Organotin Compounds in Modern Technology', Elsevier Scientific Publications, Amsterdam (1985)
99. Evans, D. W., Laughlin, R. B.: Chemosphere, *13*, 213 (1984)
100. Evans, W. H., Cardarell, N. F., Smith, D. J.: J. Toxicol. Environ. Health, *5*, 871 (1979)
101. Fanchiang, Y. T., Wood, J. M.: J. Am. Chem. Soc., *103*, 5100 (1981)
102. Farrow, B. G., Dawson, A. P.: Eur. J. Biochem., *86*, 85 (1978)
103. Fortemps, E., Amand, A., Bomboir, A., Lauwerys, R., Laterre, E. C.: Int. Arch. Occup. Environ. Health, *41*, 1 (1978)
104. Freitag, K. D., Bock, R.: Z. Anal. Chem, *270*, 337 (1974)
105. Freitag, K. D., Bock, R.: Pestic. Sci., *5*, 731 (1974)
106. Getzendaner, M. E., Corbin, H.: J. Agric. Food Chem., *20*, 881 (1972)
107. Guard, H. E., Cobet, A. B., Coleman, W. M.: Science, *213*, 770 (1981)
108. Guard, H. E., Coleman, W. M., Cobet, A. B.: Abstr. 185th Nat. Meet. Div. Environ. Chem. Am. Chem. Soc., Las Vegas, Nevada (1982)
109. Guardascione, V., Di Bosco, M.: Lav. Um., *19*, 307 (1967)
110. Hall, L. W., Pinkney, A. E.: CRR Crit. Rev. Toxicol., *14*, 159 (1984)
111. Hallas, L. E., Means, J. C., Cooney, J. J.: Science, *215*, 1505 (1982)
112. Hammann, I., Büchel, K. L., Bungarz, K., Born, L.: Pflanzenschutz — Nachr. Bayer (Engl. Edn), *31*, 61 (1978)
113. Hardon, H. J., Besemer, A. F. H., Brunik, H.: Dtsch. Lebensm. Rundsch., *58*, 349 (1962)
114. Harris, L. R., Andrews, C., Burch, D., Hampton, D., Maegerlein, S.: Rep. No. DTNSRDC/SME-78/2A, D. W. Taylor Naval Ship R and D Center, Bethesda, MD (1979)
115. Henshaw, B. G., Laidlaw, R. A., Orsler, R. J., Carey, J. K., Savory, J. B.: Recd. 1978 Ann. Convent. B.W.P.A., Cambridge, 19 (1978)
116. Heron, P. N., Sproule, J. S. G.: Indian Pulp Paper, 510 (1958)
117. Hill, R., Chapman, A. H., Samuel, A., Manners, K., Morton, G.: Int. Res. Gp. Wood Pres., Doc. No. IRG/WP/3311 (1984)
118. Hobbs, L. A., Smith, P. J.: Tin Its Uses, *131*, 10 (1982)
119. Hodge, V. F., Seidel, S. L., Goldberg, E. D.: Anal. Chem., *51*, 1256 (1979)
120. Horacek, V., Demcik, K.: Prac. Lek., *22*, 61 (1970)
121. Huber, G.: Tin Its Uses, *113*, 7 (1977)
122. Huey, C., Brinckman, F. E., Grim, S., Iverson, W. P.: Int. Conf. on Transport of Persistent Chemicals in Aquatic Ecosystems, Ottawa, 1—3 May (1974)
123. Innes, J. R. M., Ullard, B. M., Valerio, M. G., Petrucelli, L., Fishbein, L., Hart, E. R., Palotta, A. J., Bates, R. R., Falk, H. L., Gart, J. J., Klein, M., Mitchell, I., Peters, J.: J. Natl. Cancer Inst., *42*, 1101 (1969)
124. Iwai, H., Wada, O.: Indust. Health, *60*, 1 (1981)
125. Iwai, H., Wada, O., Arakawa, Y.: J. Anal. Toxicol., *5*, 300 (1982)
126. Iwai, H., Wada, O., Arakawa, Y., Ono, T.: J. Toxicol. Environ. Health, *9*, 41 (1982)

127. Jackson, J. A., Blair, W. R., Brinckman, F. E., Iverson, W. P.: Environ. Sci. Technol. *16*, 110 (1982)
128. Jeltes, R.: Ann. Occup. Hyg., *12*, 203 (1969)
129. Jewett, K. C., Brinckman, F. E.: J. Chromatogr. Sci., *19*, 583 (1981)
130. Katsumura, T.: Jap. Patent 21 360 165 (1975)
131. Keiser, D., Kanaan, A. S.: J. Phys. Chem., *73*, 4264 (1969)
131. Keiser, D., Kanaan, A. S.: J. Phys. Chem., *73*, 4264 (1969)
132. Kenaga, E. E.: U.S. Patent 3 264 177 (1966)
133. Kimmel, E. C., Casida, J. E., Fish, R. H.: J. Agric. Food Chem., *28*, 117 (1980)
134. Kimmel, E. C., Fish, R. H., Casida, J. E.: J. Agric. Food Chem., *25*, 1 (1977)
135. Klimmer, O. R.: Zentralbl. Veterinar. Med., *A11*, 29 (1964)
136. Klötzer, D.: Zentralinst. Kernforsch. Rossendorf. Dresden Zfk., *340*, 84 (1977)
137. Klötzer, D., Thust, U.: Chem. Tech. (Leipzig), *28*, 614 (1976)
138. Komora, V. F., Popl, M.: Holztech., *19*, 145 (1978)
139. Landa, K., Fejfusovo, J., Nedomlelova, R.: Prac. Lek., *25*, 391 (1973)
140. Laye, P. G.: Univ. Leeds, Unpublished work (1982)
141. Leeuwangh, P., Nijman, W., Wisser, H., Kolar, Z., de Goby, J. J. M.: Med. Fac. Landbouww. Rijkuniv. Ghent, *14*, 1483 (1976)
142. Limouzin, Y., Maire, J. C.: in 'Organotin Compounds: New Chemistry and Applications', ed. Zuckerman, S. J., Am. Chem. Soc., adv. Chem. Ser., 156, 204 (1976)
143. Luijten, J. G. A., Van der Kerk, G. J. M.: J. Appl. Chem., *11*, 35 (1961)
144. Luijten, J. G. A., Van der Kerk, G. J. M.: J. Appl. Chem., *11*, 38 (1961)
145. Lyle, W. H.: Brit. J. Ind. Med., *15*, 193 (1958)
146. McGrady, M. M., Tobias, R. S.: Inorg. Chem., *3*, 1157 (1964)
147. McGrady, M. M., Tobias, R. S.: J. Am. Chem. Soc., *87*, 1909 (1965)
148. Macchi, G., Pettine, M.: Environ. Sci. Technol., *14*, 815 (1980)
149. Maguire, R. J.: Environ. Sci. Technol., *18*, 291 (1984)
150. Maguire, R. J., Carey, J. H., Hale, E. J.: J. Agric. Food Chem., *31*, 1060 (1983)
151. Maguire, R. J., Huneault, H.: J. Chromatogr., *209*, 458 (1981)
152. Maguire, R. J., Wong, P. T. S., Rhamey, J. S.: Can. J. Fish Aquat. Sci., *41*, 537 (1984)
153. Manders, W. F., Olsen, G. J., Brinckman, F. E., Bellama, J. M.: J. Chem. Soc. Chem. Commun., 538 (1984)
154. Markicevic, A., Turko, V.: Arch. Hig. Rada. Toksikol., *18*, 355 (1967)
155. Mason, B.: 'Principles of Geochemistry', 3rd Edn., Wiley, New York (1966)
156. Massaux, F.: Café Cacao Thé, *15*, 221 (1971)
157. Meinema, H. A., Berger-Wiersma, T., Versluis-de Haan, G., Gevers, E. C.: Environ. Sci. Technol., *12*, 288 (1978)
158. Meyling, A. H., Pitchford, R. J.: Bull. WHO., *34*, 141 (1966)
159. Mijatovic, M.: Jugosl. Inostrana Dokumentacija Zastite na Rada, *8*, 3 (1972)
160. Monaghan, C. P., Hoffman, J. E., O'Brien, E. J., Frenzel, L. M., Good, M. L.: Proc. Int. Control Rel. Pestic. Symp., Corvallis, OR, 1 (1977)
161. Monaghan, C. P., Kulkarni, V. I., Ozcan, M. O., Good, M. L.: U.S. Govt. Rep., Off. Naval Res., Tech. Rep No. 2, AD-AP 87374 (1980)
162. Mor, E., Beccaria, A. M., Poggi, G.: Ann. Chem., *63*, 173 (1973)
163. Nakashima, S.: Bull. Chem. Soc. Jap., *52*, 1844 (1979)
164. Nazarenko, V. A., Antonvitch, V. P., Nevskaya, E. M.: Russ. J. Inorg. Chem., *116*, 980 (1971)
165. Neumann, W. P.: 'The Organic Chemistry of Tin', Wiley, New York (1970)
166. Odeyemi, O., Ajulo, E.: Walter Sci. Technol. *14*, 133 (1982)
167. Omar, M., Bowen, H. J. M.: J. Radioanal. Chem., *74*, 273 (1982)
168. Orsler, R. J., Holland, G. E.: Int. Biodet. Bull., *18*, 95 (1982)
169. Ozcan, M., Good, M. L.: Proc. Am. Chem. Soc. Div, Environ. Chem., Houston, Texas, March (1980)
170. Pelikan, Z., Cerny, E.: Arch. Toxicol., *27*, 79 (1970)
171. Pelikan, Z., Cerny, E.: Arch. Toxicol., *26*, 196 (1970)
172. Penninks, A. H., Seinen, W.: Vet. Qtly., *6* (4) Sept. 209 (1984)
173. Pettine, M., Millero, F. J., Macchi, G.: Anal. Chem., *53*, 1039 (1981)
174. Pettis, R. W., Philip, A. T., Woodford, J. M. D.: Austral. Def. Sci. Ser., Rep. No. 516 (1972)

175. Plum, H.: Internat. Environ. Safety, Dec., 4 pp (1982)
176. Poller, R. C.: 'The Chemistry of Organotin Compounds', Academic Press, New York (1970)
177. Prough, R. A., Stalmach, M. A., Wiebkin, P., Bridges, J. W.: Biochem. J., *196*, 763 (1981)
178. Rapsomanikis, S., Weber, J. H.: Environ. Sci. Technol., *19*, 352 (1985)
179. Reed, D. J., Woodbury, D. M., Holtzer, R. L.: Arch. Neurol., 16 (1964)
180. Rey, C., Reinecke, H. J., Besser, R.: Vet. Hum. Toxicol., *26* (2), Apr., 121 (1984)
181. Rose, M. S.: in 'Pesticide Terminal Residues', ed. Tahori, A. S., Butterworth, London (1971)
182. Rose, M. S., Aldridge, W. N.: Biochem. J. *106*, 821 (1968)
183. Rose, M. S., Lock, E. A.: Biochem. J., *120*, 151 (1970)
184. Ross, W. D.: Am. J. Psychiatry, *8*, 1092 (1981)
185. Scholz, J., Bader, C.: Lab. für Gewerbe u. Arzneimitteltoxicol. der Hoechst AG, Ber. 68 (1970)
186. Schramel, P., Sarmsahl, K., Pavlu, J.: Int. J. Environ. Studies, *5*, 37 (1973)
187. Schweinfurth, H.: Tin Its Uses, *143*, 9 (1985)
188. Seidel, S. L., Hodge, V. F., Goldberg, E. D.: Thalassia Jugoslavia, *16*, 209 (1980)
189. Seinen, W.: Vet. Sci. Commun., *3*, 279 (1979/80)
190. Seinen, W.: Immunol. Consid. Toxicol., *1*, 103 (1981)
191. Selwyn, M. J.: in 'Organotin Compounds: New Chemistry and Applications', ed. Zuckerman, J. J., Am. Chem. Soc. Adv. Chem. Ser., *156*, 204 (1976)
192. Selwyn, M. J., Dawson, A. P., Stockdale, M., Gaines, N.: Eur. J. Biochem., *14*, 120 (1970)
193. Senick, G. A.: Polymer, *23*, 1385 (1982)
194. Sheldon, A. W.: J. Paint Technol., *45*, 54 (1975)
195. Sheldon, A. W.: Proc. 18th Ann. Marine Coatings Conf. Monterey, CA., March (1978)
196. Sherman, L. R., Carlson, T. L.: J. Anal. Toxicol., *4*, 31 (1980)
197. Sherman, L. R., Yazdi, M., Hoarg, H.: Abstr. 4th Int. Conf. Organomet. Coord. Chem. Ge, Sn and Pb, Montreal, Aug (1983)
198. Skinner, H. A.: Adv. Organomet. Chem., *2*, 49 (1964)
199. Slesinger, A. E.: Proc. 17th Ann. Marine Coatings Conf. Biloxi., MO (1977)
200. Smith, P. J.: 'Toxicological Data on Organotin Compounds', Int. Tin Res. Inst., Publn. No. 538 (1978)
201. Smith, P. J.: 'Structure/Activity Relationships for Di- and Tri-Organotin Compounds', Int. Tin Res. Inst., Publn. No 569 (1978)
202. Smith, P. J., Crowe, A. J., Allen, D. W., Brooks, J. S., Formstone, R.: Chem. Ind., 874 (1977)
203. Soderquist, C. J., Crosby, D. G.: Anal. Chem., *50*, 1435 (1978)
204. Soderquist, C. J., Crosby, D. G.: J. Agric. Food Chem., *28*, 111 (1980)
205. Stein, V. T., Küster, K.: Z. Wasser Abswasser Forsch., *15*, 178 (1982)
206. Stockdale, M., Dawson, A. P., Selwyn, M. J.: Eur. J. Biochem., *15*, 342 (1970)
207. Stoner, H. B., Barnes, J. M., Duff, J. I.: Brit. J. Pharmacol., *10*, 16 (1955)
208. Suess, A., Eben, C.: Z. Pflanzenkrankh. Pflanzenschutz, *80*, 288 (1973)
209. Sutter, H. P.: M.Sc. Thesis, Portsmouth Polytechnic (1980)
210. Sylph, A.: 'Bibliography on the Neurotoxicology of Organotin Compounds', Int. Tin Res. Inst., Publn No LB 12 (1984)
211. Sylph, A.: 'Bibliography on the Toxicity of Organotins to Aquatic Animals', Int. Tin Res. Inst., Publn. No. LB 11 (1984)
212. Taketa, F., Siebenlist, K., Kasten-Jolly, J., Palosaari, N.: Archiv. Biochem. Biophys., *203*, 466 (1980)
213. Tan, L. P., Ng, M. L., Kumar Das, V. G.: J. Neurochem., *31*, 1035 (1978)
214. Thayer, J. S.: in 'Organometals and Organometalloids: Occurrence and Fate in the Environment', eds. Brinckman, F. E., Bellama, J. M., Am. Chem. Soc. Symp. Ser., *82*, 188 (1978)
215. Thayer, J. S.: Abstr. 10th Int. Conf. Organomet. Chem., Toronto, Aug (1981)
216. Thayer, J. S.: Paper presented to the 4th Int. Conf. Organomet. Coord. Chem. Ge, Sn and Pb, Montreal (1982)
217. Thayer, J. S., Olsen, G. J., Brinckman, F. E.: Environ. Sci. Technol., *18*, 726 (1984)
218. Tobias, R. S.: in 'Organometals and Organometalloids: Occurrence and Fate in the Environment', eds. Brinckman, F. E., Bellama, J. M., Am. Chem. Soc. Symp. Ser., *82*, 130 (1978)
219. Tominaga, M., Umerzaki, Y.: Anal. Chim. Acta, *110*, 55 (1979)
220. Thust, U.: Tin Its Uses, *122*, 3 (1979)
221. Tugrul, S., Balkas, T. I., Goldberg, E. D., Salikoglu, I.: J. Etud. Pollutions, 497 (1982)

222. Tugrul, S., Balkas, T. I., Goldberg, E. D.: Mar. Pollut. Bull., *14*, 297 (1983)
223. Tzschach, A., Ponicke, K.: Kem. Kozlem., *41*, 141 (1974)
224. Underwood, E. J.: 'Trace Elements in Human and Animal Nutrition', Academic Press, New York, (1971)
225. Van der Kerk, G. J. M., Luijten, J. G. A.: J. Appl. Chem., *4*, 301 (1954)
226. Van der Kerk, G. J. M., Luijten, J. G. A.: J. Appl. Chem., *4*, 314 (1954)
227. Van der Kerk, G. J. M., Luijten, G. A.: J. Appl. Chem., *6*, 49 (1956)
228. Van der Kerk, G. J. M., Noltes, J. G.: J. Appl. Chem., *7*, 369 (1957)
229. Van der Kerk, G. J. M., Noltes, J. G.: J. Appl. Chem., *9*, 106 (1959)
230. Van der Kerk, G. J. M., Noltes, J. G., Luijten, J. G. A.: J. Appl. Chem., *7*, 356 (1957)
231. Van der Kerk, G. J. M., Noltes, J. G., Luijten, J. G. A.: J. Appl. Chem., *7*, 366 (1957)
232. Vind, H. P., Hochman, H.: Proc. Am. Wood Pres. Assoc., *58*, 170 (1962)
233. Vinogradov, A. P.: Sears Foundation for Mar. Res., New Haven, Yale Univ., 16 (1953)
234. Wada, O., Manabe, S., Iwai, H., Arakawa, Y.: Jap. J. Ind. Health, *24*, 24 (1982)
235. Waldock, M. J., Miller, D.: Int. Council on the Exploration of the Sea, Rep. No. CM1983/E:12, Copenhagen, (1983)
236. Waldock, M. J., Thain, J., Miller, D.: Int. Council on the Exploration of the Sea, Rep. No CM1983/E:52, Copenhagen (1983)
237. Waldock, M. J., Thain, J. E.: Mar. Pollut. Bull., *14*, 411 (1983)
238. Ward, G. S., Cram, G. C., Parrish, P. R., Trachman, H., Slesinger, A.: Aquat. Toxicol. Haz. Assess: 4th Conf., ASTM STP 737, ed. Branson, D. R., Dickson, K. L., Am. Soc. Testing and Materials, p 183 (1981)
239. Watkins, D. A. M.: Chem. Ind., 185 (1974)
240. Watling-Payne, A. S., Selwyn, M. J.: Biochem. J., *142*, 65 (1974)
241. Woggon, H., Jehle, D.: Die Nahrung, *19*, 271 (1975)
242. Yu, T., Arakawa, Y.: J. Chromatogr., *258*, 189 (1983)
243. Zimmerli, B., Zimmerman, H.: Fresenius Z. Anal. Chem., *304*, 23 (1980)

Chemicals Used in the Rubber Industry

*Lawrence Fishbein**

ENVIRON Corporation Washington, D.C., USA

Summary

A large number of chemicals illustrative of many structural and use categories are employed in the rubber industry. This chapter surveys the major rubber chemicals as to their utility, production trends, structural characteristics, the nature of potential impurities, toxicology (with primary focus on carcinogenicity and genotoxicity) as well as areas of specific exposure, concern, and epidemiology in the rubber industry. Additionally, the disposal and reclamation of rubber as well as the emission, degradation and analysis of rubber chemicals are considered.

Introduction

A broad spectrum of chemicals numbering into the hundreds and encompassing many structural and use categories are employed in the rubber industry for a multitude of applications. Additives can thus be classified according to their functions which include: vulcanizing agents, curing agents, accelerators, antidegredants (antiozonants, antioxidants), activators, retarders, blowing agents, processing aids, reinforcing agents. mould release agents, plasticizers, solvents, bonding agents and fillers [1—3]. It has been reported that some 500 chemically, more or less, well-defined

---

* Current address: ILSI — Risk Science Institute, Washington, D.C., USA

substances are used in quantities ranging from 10 to 10^6 kg annually in the Swedish rubber industry alone [2, 4].

Initially it is useful to examine the historical development of the rubber industry and the elastomers and use products as it bears on the type and amount of chemicals employed and the potential populations at risk.

Historical Overview

A number of early developments have spurred the growth of the rubber industry since the initial discovery of vulcanization by Goodyear in 1844 and the growth and global spread of rubber plantations. These include the development by 1910 of the motor car in the United States and the consequent burgeoning demand for rubber followed the rapid expansion of industrial applications of rubber [1—3]. With few exceptions, raw rubber in the dry state has few commercial applications. For the great majority of uses, the rubber must be modified, usually by the addition of vulcanization agents and other materials.

The vulcanization, or curing process, is based on the reaction of unsaturated rubber hydrocarbon molecules with sulphur in the presence of zinc oxide and other inorganic bases, continued to be used during the latter half of the nineteenth and the early years of the twentieth century. During the first quarter of the twentieth century the bulk of all rubber polymer used in manufacturing processes consisted of natural rubber [3]. Although patents for synthetic polymer were first taken out in 1912, it was not till 1930 that the development of synthetic polymer was intensified in Germany and the USSR [3]. A number of synthetic rubbers were in commercial production in Europe by 1939, in the United States by 1944 and in the United Kingdom by the mid-1950s [3].

There were only four main types of synthetic rubber available in the mid-1950s. These were: styrene-butadiene rubber (SBR), butyl, nitrile and polychloroprene rubbers. Advances in the synthesis of elastomers in the 1960s produced polyisoprene, polybutadiene and ethylene-propylene terpolymer (EPDM).

Recent Production Trends

The development and usage of synthetic polymer is considered to be the most important of all the process changes that have taken place in the rubber industry in recent years [3].

Additional rubbers that are now available include: chlorohydrin elastomers, polysulfides, polyacrylates, halogenated butyl rubbers, silicone rubbers, polyurethanes, chlorinated polyethylene, fluoroelastomers, and thermoplastic elastomers and ethylene-propylene terpolymer (EPDM).

Figure 1 illustrates the structures of some major elastomers [6] and Table 1 lists the natural and 32 synthetic elastomers in use [3].

The dramatic use in the production of synthetic rubbers can be gleaned from the following considerations. In 1955, the U.S. and Canada manufactured most of the *synthetic* rubber in the Western world with smaller amounts being produced in the

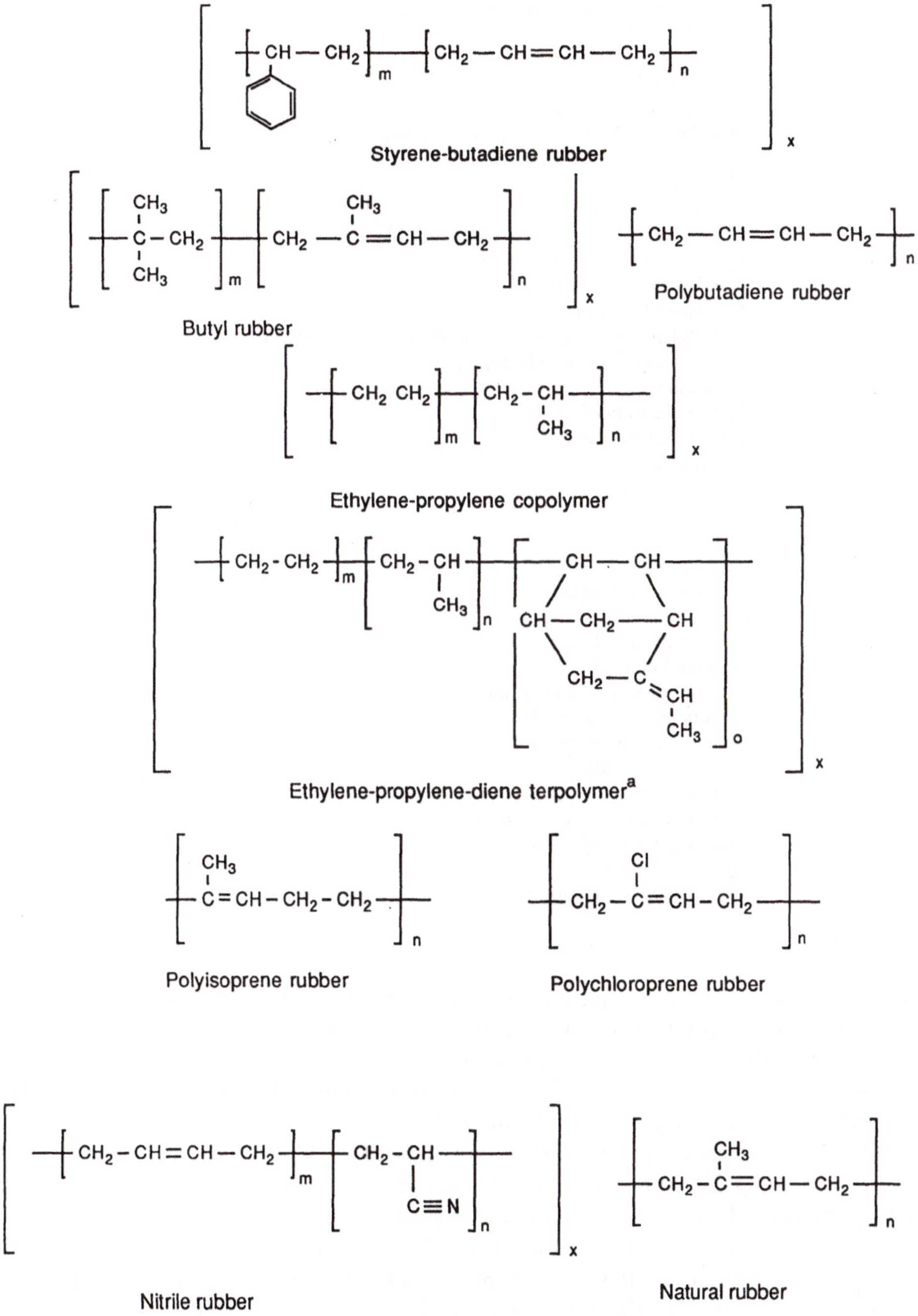

Styrene-butadiene rubber

Butyl rubber

Polybutadiene rubber

Ethylene-propylene copolymer

Ethylene-propylene-diene terpolymer[a]

Polyisoprene rubber

Polychloroprene rubber

Nitrile rubber

Natural rubber

[a]Third monomer shown is dicyclopentadiene ethylidene norbornene and 1,4-hexadiene are also used

Fig. 1. Structures of major elastomers

Table 1. Natural and synthetic rubber elastomers

Natural rubber
Synthetic rubbers:
  Acrylic rubbers (NBR)
  Acrylonitrile-butadiene-styrene copolymer
  Chlorinated polyethylene
  Chlorobutyl rubber
  Chloroprene rubber
  Chlorosulphonated polyethylene (CSM)
  Epichlorohydrin elastomer (CO)
  Ethylene-propylene-diene terpolymer (EDPM)
  Fluoroelastomers (CFM)
  Isobutylene-isoprene elastomers (butyl rubber) (IIR)
  Isoprene-acrylonitrile elastomers
  Neoprene
  Nitrile elastomers (NBR)
  Polybutadienes (BR)
  Polybutenes
  Polychloroprenes (CR)
  Polyester
  Polyethylene
  Polyisobutylenes
  Polyisoprenes (MR)
  Polysulphides (T)
  Polytetrafluoroethene
  Polyvinyl chloride
  Resorcinol-formaldehyde latex
  Silicone elastomer (MQ)
  Styrene-butadiene rubbers (SBR)
  Oil-extended SBR
  Styrene-isoprene elastomers
  Thermoplastic polyester elastomers
  Thermoplastic polyolefin elastomers
  Urethane rubbers
  Vinyl acetate-ethylene copolymers

Federal Republic of Germany, the German Democratic Republic and the USSR. While worldwide production at that time was only 1,542,000 tonnes, by 1978 worldwide production had grown to 8,720,000 tonnes, a growth rate of 7.8% per year. During the same period 1955–1978, the use of natural rubber grew from 1,928,000 tonnes to 3,725,000 tonnes, a growth rate of 2.9% per year [5].

The 5 year forecast (1988–1992) for worldwide rubber consumption calls for growth averaging 2% annually. By 1992, worldwide consumption of synthetic and natural rubber will be 15.9 million metric tons which represents an 11% rise from the 14.4 million metric tons estimated for 1987. The consumption for 1982 is estimated to have risen 3.3% over 1986's 13.9 million metric tons. About 68% of all new rubber consumed worldwide in 1987 was synthetic. The 1987 global consumption of synthetic rubber, including communist countries, increased 3.1% from 1986 to 9.75 million metric tons. In the U.S. and Canada, combined consumption increased 3.3% to 2.53 million metric tons in 1987. Synthetic rubber demand is forecast to continue growing in these two countries, but only modestly for a 1.5% increase over the next 5 years to 2.56 million metric tons in 1992. Table 2 illustrates the forecast for increases in synthetic and natural rubber from 1987 to 1992 for U.S. and Canada, Western Europe, Latin America, Asia and Oceania and Africa and the Middle East [6].

Table 2. Forecast for Synthetic and Natural Rubber Use from 1987 to 1992

Thousands of metric tons	U.S. & Canada		Western Europe		Latin America		Asia & Oceania		Africa & Middle East		Total	
	1987	1992	1987	1992	1987	1992	1987	1992	1987	1992	1987	1992
Styrene-butadiene[a]	839	833	727	745	329	386	674	783	63	72	2,832	2,819
Carboxylated styrene-butadiene	447	458	335	370	21	29	125	152	5	6	933	1,015
Polybutadiene	422	424	240	257	109	131	270	309	19	21	1,060	1,142
Ethylene-propylene diene	204	216	170	195	11	16	107	126	2	3	494	556
Polychloroprene	86	85	81	77	19	23	68	78	5	5	259	268
Nitrile	68	66	81	85	13	17	51	66	3	4	216	238
Other synthetics[b]	457	479	216	244	39	46	197	223	22	33	931	1,025
TOTAL SYNTHETIC	2523	2561	1850	1973	541	618	1492	1737	119	144	6,525	7,063
Natural	872	871	945	980	247	296	1568	1804	140	180	3,771	4,131
TOTAL RUBBER	3395	3432	2460	2953	788	944	3060	3541	259	324	10,296	11,194

Note: Excludes communist countries. a Solid and latex forms of SBR. b Includes polyisoprene and butyl. Source: International Institute of Synthetic Rubber Productions.

Uses of Synthetic and Natural Rubbers

Tires continue to dominate rubber use as has been the case for many decades. Roughly half of all rubber consumption goes into the production of tires [6]. In non-communist countries, on a weighted average basis, tires and tire products, consume 43% of all synthetic rubber and 62% of natural rubber. In the U.S. and Canada, tires and tire products account for 43% of synthetic rubber consumption and 72% of natural rubber consumption. In Western Europe, 35% of synthetic rubber and 66% of natural rubber go into tires. In Latin America, tires and tire products account for the largest shares (62% of synthetic and 77% of natural) rubber consumption among areas of the world [6].

a) *Styrene-butadiene Rubber* (SBR)

In regard to the individual classes of synthetic rubbers, tires and tire components such as tread rubber consume more than 70% of styrene-butadiene rubber (SBR) produced the U.S. [6, 7]. Non-tire uses of SBR account for about 25% of consumption. A large share of these users are in automotive applications (where SBR often is blended with other elastomers), such as hoses, belts, gaskets, seals and various external or moulded items [6]. Additional non-tire and non-automotive users of SBR range over most industries that require hoses, belts, gaskets, seals and other external and moulded products. Significant amounts of SBR are also utilized in footwear, various kinds of solid wheels, roll covers, coated fabrics, and wire and cable insulation. A sizeable amount of SBR is sold as the latex to be applied as carpet backing [6, 7].

b) *Polybutadiene*

In the U.S., solution-polymerized *cis*-1,4-polybutadiene is second to SBR in share of synthetic rubber production. In North America, more than 80% of polybutadiene is used in tires while worldwide, tires and tire products consume more than 77% of the total polybutadiene output. Polybutadiene's major tire use is in treads for which it provides long wear and less resistance to rolling than most lower-cost elastomers. Polybutadiene is often blended with SBR and natural rubber in varying amounts depending on the nature of the tire being made. While passenger car tire treads generally are blends of polybutadiene and SBR, truck and bus tire treads are blends of polybutadiene and natural rubber [6, 7].

About 20% of the U.S. consumption of polybutadiene is as an impact modifier for polystyrene and other resins by graft polymerization. Impact-modified polystyrenes include polybutadiene contents in the range of 2 to 10%. The remainder of poly-butadiene elastomer consumption is in various extruded and moulded industrial rubber products.

c) *Ethylene-propylene Elastomers*

In terms of production, ethylene-propylene elastomers rank next to polybutadiene. The copolymers are produced in much lower volumes than are the terpolymers, with terpolymers accounting for three-quarters of U.S. output. The third type of mono-mer, known as dienes, in terpolymer of commercial significance is either dicyclo-pentadiene, ethylidene norbornene, or 1,4-hexadiene. Each of these termonomers yield different characteristics in the final elastomer. For example, dicyclopentadiene yields elastomers with the slowest cure rates of the dienes and yields a polymer with branched chains. Ethylidene norbornene produced by recycling butadiene with cyclobutadiene may be the most extensively used of the termonomers since it is easily incorporated into the polymer. The double bond so introduced improves sulfur vulcanization [6].

A wide range of compositions is possible for the co- and terpolymers of ethylene-propylene. Most commercial grades of copolymer elastomers have 40 to 60% of ethylene in them while commercial grades of terpolymers (EPDM) have about the same ratio of ethylene to propylene and up to about 10% of diene. Currently the automotive market (excluding lubricating oil additives) is still the largest use for ethylene-propylene elastomers. Tire use for ethylene-propylene elastomer is small with automotive uses predominantly encompassing hoses, seals, grommets and weather stripping. Additional amounts of ethylene-propylene elastomers are blended with polyolefins to make thermoplastic olefin elastomers that are used in bumper parts, some exterior body parts (e.q., fender extensions and rub strips), wire and cable coverings and numerous other items. Non-automotive uses of ethylene-propylene elastomers include: single-ply roofing, wire and cable insulation, hoses of many kinds, appliance parts and other extruded and moulded products [6, 8].

d) *Polyisoprene*

Polyisoprene is made by additive polymerization (analogous to the process for polybutadiene). Because of the 1,3-dienes structure of isoprene, the reaction can involve the same three double bond positions as in butadiene and can yield different configurations. Of the eight possible arrangements for polyisoprene, three have been isolated: the *cis*-1,4, *trans*-1,4-, and atactic-3,4-forms. Because of its similarity to natural rubber, the cis-1,4-form of polyisoprene is used principally in the production of tires mainly in sidewalls for passenger cars and light truck tires). More polyisoprene, as a share of total rubber content, goes into truck and bus tires where it substituted for natural rubber in blends with polybutadiene to provide the properties of good resistance to heat build up and good abrasion resistance in treads [6].

Non-tire uses of polyisoprene elastomers account for about 40% of consumption and include products such as mechanical goods (made by extrusion or moulding), for automotive and other industries, footwear, and many other smaller-volume items [6]. In many of these uses, these elastomers substitute for natural rubber. It should be noted that the capacity for making polyisoprene elastomers in the USSR is estimated to be more than 1 million metric tons per year which is several times the combined capacity in noncommunist countries. Such large capacity is intended to minimize dependence on natural rubber, now slightly more than 20% of total new rubber consumption in communist countries [6].

e) *Butyl Rubber*

Butyl rubbers, including the halobutyl rubbers, are also important tire elastomers being extensively employed as the inner liner for tubeless tires and in tubes for tires. Butyl elastomers are copolymers of isobutylene and a small amount isoprene. The commercial halobutyls are either chloro or bromo-derivitives. Besides ozone and weather resistence, butyl elastomers can be tailored to have good thermal, chemical, moisture and vibrational resistance. Non tire uses of butyl elastomers include sealants and adhesives, hoses, gaskets, pads for truck cabs, bridge bearing mounts [6].

f) *Polychloroprene Elastomers*

Polychloroprene elastomers are polymers of chloroprene, 2-chloro-1,3-butadiene that are made by emulsion polymerization. Variations in the polymerization process lead to a large variety of elastomers, both solid and latex forms, designed for specific applications. Polychloroprene elastomers are generally noted for their resistance to abrasion, hydrocarbons, sunlight, ozone and weathering and their toughness. The major uses for polychloroprene elastomers are in making mechanical rubber goods for automotive, petroleum production transportation, construction and consumer products. The elastomers are fabricated into well packers, other seals, gaskets and hoses while polychloroprene lattices are used in making gloves, adhesives and binders. Small amounts of polychloroprene elastomers are used in some tire sidewalls but are not considered a tire rubber [6—8].

g) *Nitrile Elastomers*

Nitrile elastomers are copolymers of acrylonitrile and butadiene with properties varying principally depending on the composition of the polymers. In general, the higher the acrylonitrile content (40–50%), the higher the rubber's resistance to hydrocarbons, abrasion and gas permeation. Rubber with low acrylonitrile content (about 20%) are more resilient and retain resilience at lower temperatures. The polymerization process for nitrile rubbers are similar to those used for emulsion SBR and are operational either as hot or cold and as batch or continuous processes. Extensive compounding is required to give fabricated nitrile rubber products adequate resistance to ozone and oxidants. Because of their hydrocarbon resistant property, nitrile rubbers are principally employed in the form of hoses, tubing, gaskets, seals and packers to handle oils, fuels and chemicals [6, 7]. The above fabricated items from nitrile elastomers are used in diverse transportation equipment, food processing and crude petroleum production items. Additionally various nitrile elastomers are also used to modify plastics such as polyvinyl chloride (PVC) and acrylonitrile-butadiene-styrene (ABS) to impart impact resistance. Uses of the latex form include paper saturation for masking tapes, binding papers and label stock [6, 7].

h) *Specialty Elastomers*

A number of important specialty elastomers are produced which possess a wide variety of physical properties. The specialty rubbers are more expensive than the large-volume general-purpose rubbers. In the U.S. all of the specialty rubbers do not exceed 100,000 metric tons or less than 5% of the total synthetic rubber production. Of the specialty rubbers, silicone and chlorosulfonated polyethylene could account for approximately half of the total specialty production [7, 8]. Silicone, urethane and fluorine-based elastomers are among the fastest-growing products of the synthetic rubber industry [6, 7].

Chlorosulfonated Polyethylenes

These rubbers are prepared from crystalline polyethylene that is modified by chlorine and sulfonyl groups. They possess toughness, high resistance to ozone and light, good resistance to oxidation, heat and oil and corrosive materials and are employed in wire and cable coverings, roofing, automotive parts, pond liners, hoses and belting [7]. It was estimated that approximately 60 million pounds of these rubbers were used in the U.S. in 1986.

Silicones

Silicone elastomers based on polymers of dimethylsiloxane, possess high resistance to ozone, weathering and hydrocarbons. Various side groups (e.g., vinyl or phenyl), which can be added to the basic silicon-oxygen chain impart other properties. Most of the silicone elastomers are employed as sealants and adhesives with smaller quantities used in automotive parts, wire and cable coverings and medical equipment [7, 8]. Approximately 75 million pounds of silicone elastomers were estimated to be used in 1986 in the U.S.

Acrylics

Acrylic elastomers are principally made of ethyl-, butyl-, or methoxy and ethoxyethyl acrylates. Their major uses are in seals, gaskets and hoses. The U.S. annual consumption of acrylic elastomers is approximately 10 million pounds [6, 7].

Epichlorohydrins

Both homo- and copolymers of epichlorohydrin are fabricated into elastomeric products that possess good resistance to petroleum products, and have low gas permeability and low-temperature flexibility. These elastomers are mainly used in automotive products such as gaskets, hoses and motor mounts. Approximately 15 million pounds of epichlorohydrin elastomers were consumed in the U.S. in 1986 [6, 7].

Fluorinated Polymers

Fluorinated elastomers are the most expensive of the specialty thermosets and are mainly copolymers of vinylidene fluoride and hexafluoropropylene, or other C_2 and C_3 fluoro-olefins. The fluorinated polymers have good resistance to heat, most solvents, and weathering as well as possessing good mechanical properties. The elastomers are principally used as seals, gaskets, hoses and other items in automobiles, oil-well drilling and other industries [7]. Approximately 5 million pounds of fluorinated elastomers are consumed annually in the U.S.

Polyurethanes

The polyurethanes have become the most widely used among the specialty elastomers. Estimates of total production are at more than 200,000 metric tons per year [6]. These elastomers can be made from a variety of monomers by solution, suspension and melt polymerization. The polymers that are fabricated by casting, moulding by different methods, milling and other means are used as seals, gaskets, automotive body parts, solid tires, many kinds of hoses and cable coverings as well as in consumer products such as in seating. Additionally, polyurethanes are employed in adhesives and fibers [6, 7].

Polysulfides

Polysulfides represent the oldest of the specialty elastomers having been introduced more than 50 years ago. Approximately 20 million pounds of polysulfides were consumed in 1986, largely as sealants. These elastomers possees good resistance to ozone, oxygen, and sunlight and have good low temperature properties and low permeability to gases and liquids [6, 7].

Copolyester Ethers

Copolyester ether elastomers result from the condensation of polybutylene terphthalates and polytetramethylene etherglycol terephthalate. They possess high strength, abrasion resistance, a relatively wide temperature range and good resistance to chemicals and petroleum. Approximately 20 million pounds of these elastomers were estimated to be used in 1986 in the U.S. as hose, wire and cable coverings, as well as in various automotive parts [7].

Thermoplastic Elastomers (TPE)

Thermoplastic elastomers (TPE) are intermediate between rubbers and plastics. At moderate ambient temperatures they have the properties of vulcanized rubber, whereas at elevated temperatures they are melt-processable like thermoplastics. TPE's can be categorized into at least six broad areas: (a) styrenics (styrene-butydiene block conlymers; styrene-butylene; styrene-ethylene-propylene, styrene-isoprene); (b) polyolefins; (c) elastomeric alloys; (d) polyurethanes, (e) copolyesters and (f) polyamides. The development of TPE's in the 1950's represented a major advance in elastomer technology. Thermoplastic resins flow under heat and pressure into a mould, are cross-linked and become permanently set. Scrap can be recovered and in most cases, re-run. This aids in eliminating the need for compounding and a longer cure cycle also helps reduce costs. Because of such characteristics the thermoplastic elastomers have been increasingly employed into various automotive and other applications. The growth rate of the TPE's is estimated at 6—7% annually [8].

Rubber-Processing Chemicals-General Considerations

The *commercially significant rubber-processing* chemicals comprise above 125 different organic chemicals and are produced in the U.S. by about 20 large and small companies [6]. These chemicals are only part of the very large quantities of materials which are employed in the production of a large variety of rubber products. Other materials which are employed include: inorganic chemicals such as pigments and fillers; minerals such as sulfur; metals; oils; fibers of both organic and inorganic compounds and reinforcing agents, such as carbon black [1, 3, 9, 10].

Before processing, raw rubber has few commercial applications. Compounding rubbers with various chemicals and subjecting the fabricated forms to heat and pressure converts them into useful items such as tires, hoses and gaskets. For example, natural rubber requires that more processing chemicals be used to lower its viscosity during fabrication, but much of that requirement for chemicals is offset by a natural antioxidant that is protective during fabrication thereby reducing the requirement for synthetic antioxidants during fabrication.

Thus in general, the aim of the rubber compounder is to formulate a composition which will be: 1) easy to put through such processes as calendering and extrusion with the minimum of energy input; 2) stable at the temperatures generally encountered in those processes but which will cure rapidly and controllably at vulcanization temperatures; and 3) capable of yielding a product processing the derived physical characteristics and which is resistant to oxidation, heat, aging and attack by light, ozone and other chemicals [3].

Hundreds of different rubber chemicals may be used in different blends, however a dozen different components may be used [1–3, 6, 11]. On the average between 4.5 and 5.0 pounds of rubber-processing chemicals are used per 100 pounds of rubber during fabrication. Much higher quantities of these chemicals will be used for some rubber products because of the severe conditions to which the product will be subjected. Hence, the variety of chemicals used in formulation depends on the product service demands [3, 6].

Classification of Rubber Chemicals and Their Production Volumes

The wide spectrum of rubber additives can be classified either by their functional activity during or after rubber processing or by their structural type [1–3, 6].

a) *Accelerators of Vulcanization*

This group of rubber additives, catalyzes the vulcanization (or curing) process by aiding the cross-linking of rubber polymer chains with sulfur.

The development and use of accelerators has greatly increased efficiency in rubber processing and fabrication. The most common vulcanization agent is sulfur which is used in most general purpose material and SBR systems. This results in the formation of cross-links and cyclic structures which changes the rubber from a soft thermoplastic mixture to a final product with the desired properties [3, 6, 10].
Sulfur donors such as tetralkylthiuram disulphides, morpholine disulphide, dithiocarbamates and, more recently, dithiophosphates are at times used instead of or in addition to elemental sulfur. Additionally, the closely related elements selenium and tellurium and their compounds have also been employed as vulcanization agents [1–3, 10].

The use of sulfur alone to vulcanize rubber requires long reaction times. Shortening the vulcanizing process permits the addition of materials that provide desired properties but that would be completely destroyed during long curing times. Since the reaction between sulfur and rubber is slow, accelerators are employed to obtain shorter cure times and better final properties. Accelerators function at normal curing temperatures (140–200 °C) and (exception in latex) not at lower temperatures. Accelerators can be classified in various ways: (a) classification by sulfur demand; less active accelerators require a relatively large amount while more active accelerators require a smaller quantity; (b) classification by rate of vulcanization achievable; slow accelerators include amines and thiourea derivatives; semi-ultra accelerators (moderately fast) include sulphenamides, 1,3-diphenylguanidine and mercaptothiazole; and ultra accelerators (very fast) include thiurams, dithiocarbamates, xanthates and thiophosphates; and (c) classification by chemical composition. The main structural groups employed as accelerators are shown in figure 2 and include aldehyde-amine condensates, dithiocarbamates, benzothiazoles, amines, thiophosphates, guanidines, thioureas, xanthates, sulfenamides and thiuram

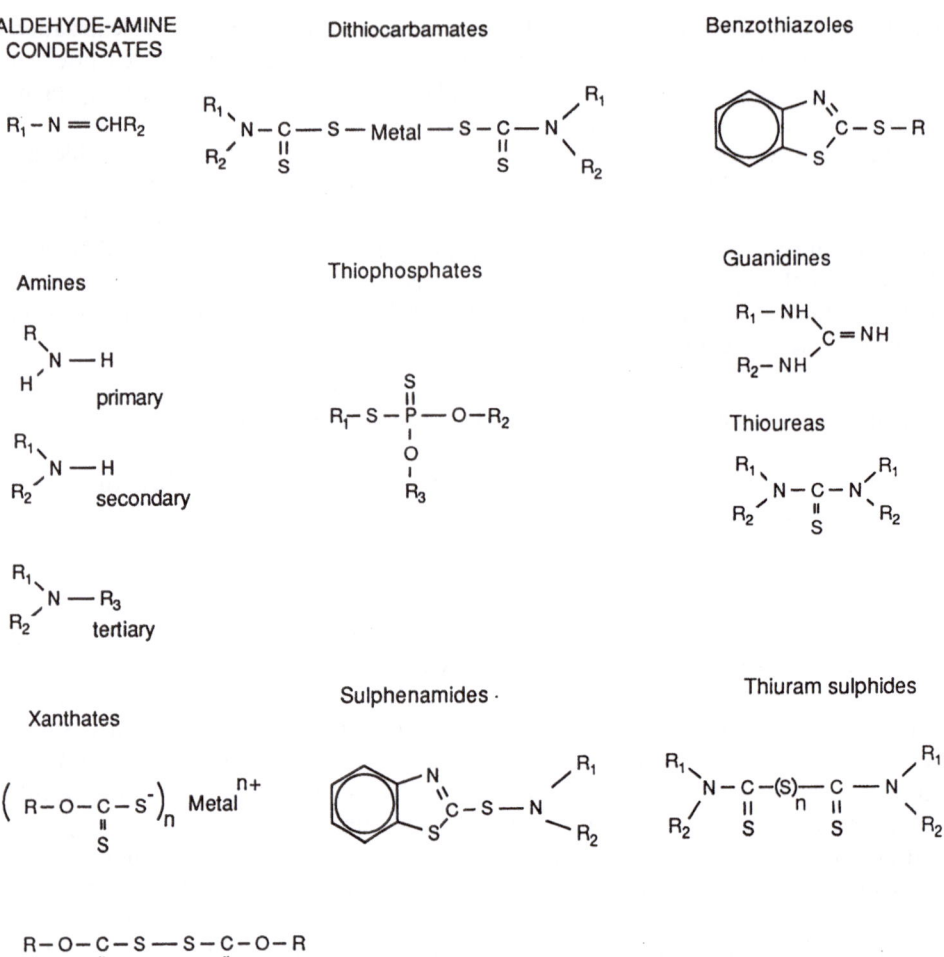

Fig. 2. Accelerator types classified by chemical structure

sulfides [3]. Typical accelerators from the above mentioned categories include (i) aldehyde-amine reaction products (e.g., acetaldehyde/ammonia, acetaldehyde/aniline, formaldehyde-p-toluidine); (ii) arylguanidines (e.g., diphenylguanidine, triphenylguanidine, di-o-tolyguanidine); (iii) dithiocarbamates (e.g., copper, lead, bismuth, zinc and selenium, tellurium, and cadmium diethyldithiocarbamates), thiuram sulfides [(e.g., tetramethyl- and tetrabutyl thiuram monosulfide, tetramethyl and tetraethyl thiuram disulfide), (v) thiazoles [e.g., 2-mercaptobenzothiazole, zinc benzothiazolylmercaptide, 2,2'-dithio-bisbenzothizole]; (vi) sulfenamide (e.g. N-cyclophenyl-2-benzothiazole- and N-oxydiethylene-2-benzothiazole sulfenamides and (vii) miscellaneous [e.g., trimethyl thiourea, 1,3-diethylthiourea, and 1,3-bis-(2-benzothiazolymercaptomethyl)-urea] [10].

It should be noted that accelerators can often behave differently in different polymer systems and that a primary accelerator may be used together with a smaller amount of secondary accelerator to produce a synergistic effect giving better final properties than can either accelerator separately [3, 6, 10]. Figure 3 illustrates the catalytic action of accelerators with rubber leading to the formation of sulfide cross-links.

The first accelerators were metal oxides e.g., lead oxide, calcium oxide and magnesium oxide. The first organic accelerator was aniline which was first used for that purpose early in the 20th century. Although aniline is no longer used because of its toxicity and technical reasons, derivatives of aniline are employed.

The discovery of 2-mercaptobenzothiozole more than 60 years ago as an efficient vulcanization accelerator spawned the development of various delayed-action accelerators used today for high output rates of rubber products [6].

Approximately, 43,000 metric tons of organic vulcanization accelerators was produced in the U.S. in 1980 [10] with a large portion of this volume comprised of derivatives of 2-mercaptobenzothiazole (MBT) such as sulfenamides.

The production of accelerators in the U.S. in 1986 amounted to 65 million pounds of which sulfenamides accounted for 50%, benzothiazoles, 22%, dithiocarbamates, 5%, thiurams, 4% and all others, 19% [6]. Of the approximately 80 million pounds of accelerators used in 1987, nearly half were projected to be sulfenamides, benzothiazoles (16–18 million pounds), dithiocarbamates (less than million pounds), and thiurams (about 3 million pounds). A large number of other compounds, many containing nitrogen and blends of accelerator compounds totalled about 15 million pounds [6].

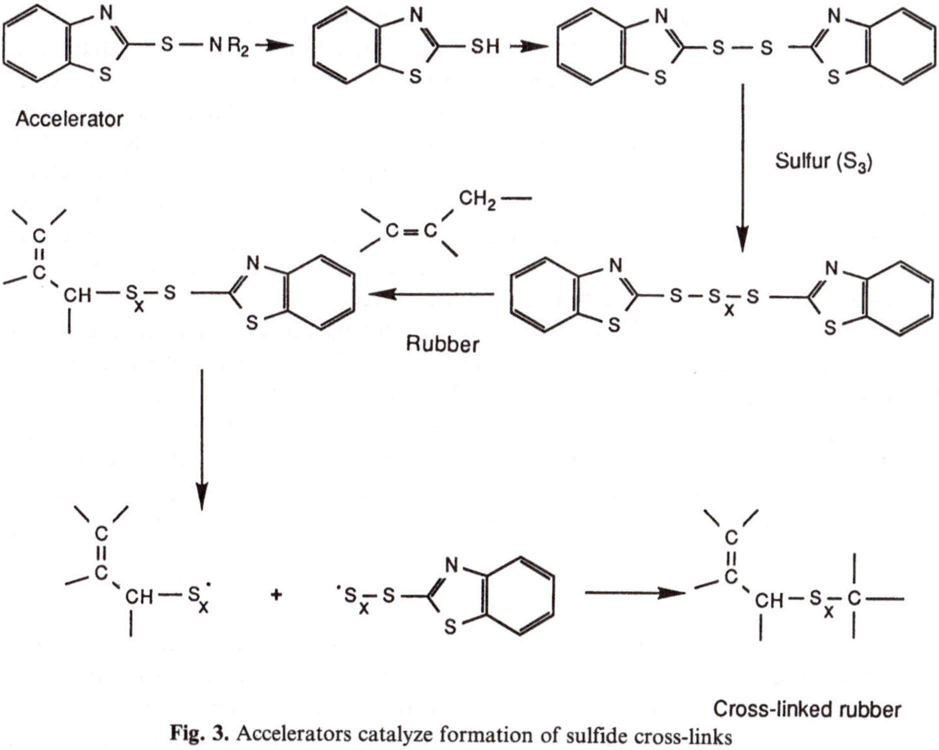

Fig. 3. Accelerators catalyze formation of sulfide cross-links

b) *Activators*

Activators are also used to make accelerators more effective by forming inter-mediate complexes. The principal types of activators are inorganic compounds (mainly metal oxides) such as zinc oxide, litharge (PbO), red lead (Pb_3O), magnesium oxide and sodium carbonate. Organic acids (e.g., stearic acid or lauric acid) are used to increase the solubility of the metal in the rubber formulation. The most common activation system is zinc oxide in combination with a fatty acid which produces a rubber soluble soap (zinc stearate) in the rubber matrix [3].

c) *Peroxide Curing Agents*

Although dibenzoyl peroxide was discovered in 1914 to cross-link rubber, the commercial use of the more effective dialkyl peroxides started shortly after 1950 [10] Dialkyl peroxides cure rubber via a free radical mechanism (e.g., the homolytic decomposition of the peroxide produces alkoxy radicals). These radicals remove a hydrogen atom from a polymer chain yielding an alkyl radical on the polymer chain. The radicals on the polymer chains join to form cross-linkages [9, 10]. Since the carbon-to-carbon bonds in these cross-links have higher bond energies than do the sulfur-sulfur and sulfur-carbon bonds of vulcanizations that involve sulfur, the products are more thermally stable than sulfur-cured vulcanizates. Peroxide cures are generally more uniform than sulfur cures with the resultant fabricated products possessing greater heat resistance and low compression set. However, a number of disadvantages of peroxide cures should be noted which include: a) peroxide cures cause different effects on different rubber polymers; b) there is the requirement for the peroxide to decompose as completely as possible before the polymer mixture is exposed to oxygen; and c) peroxides can react with other chemicals such as anti-degradants added to the rubber [9]. It should also be noted that rubber polymers with high amounts of hydrogen tertiary carbon atoms (e.g., butyl rubber) are less suitable for peroxide cures since radicals on chains are liable to cause chain breakage when cross-linking occurs. The annual volume of alkyl peroxides used in rubber compounding is comparatively small (estimated at 1.5 million pounds) with a large share of that volume due to one agent, dicumyl peroxide [9]. However, a variety of other peroxides have also been employed.

Peroxide cross-linking efficiency can also be improved and vulcanizate properties modified by certain additives. A variety of polyfunctional active olefins have been found effective (e.g., ethylene glycol dimethacrylate; trimethylolpropane trimetha-crylate, triallyl cyanurate, zinc acrylate, N,N'-m-phenylenedimaleimide and poly-butadiene). They appear to function via rapid polymerization of the active double bonds to yield a second polymer network which can be initiated by an alkyl radical from the peroxide or from a radical on the base polymer [10].

d) *Retarders*

Retarders are chemicals that prevent the premature vulcanization of rubber com-pounds during mixing, calendering and other processing steps. These chemicals are also known as antiscorching agents, cure retarders and prevulcanization inhi-bitors. They are principally used to delay the action of an accelerator and hence to prevent premature curing during intermediate processing operations or during storage of highly accelerated rubber compounds [3, 9, 10].

Conventional retarders have included benzoic acid, salicyclic acid, phthalic anhydride and N-nitrosodiphenylamine (NDDA). A more recently introduced retarder in the early 1980's, N-(cyclohexylthio)-phthalimide (CTP) (which is converted to phthalimide during vulcanization) is the principal retarder employed. In the early 1980's the total annual consumption of these retarders in the U.S. was about 1800–2300 tons, of which N-(cyclohexylthio)-phthalimide (CTP) accounted for about 100 tons [10]. Of this, about 80% of the retarders were employed in the manufacture of tires and the remainder in miscellaneous mechanical goods, footwear and sheet and calendered goods [10].

NDDA was widely used as a retarder for many decades. Its production in the U.S. peaked in 1974 reaching about 1600 metric tons and then gradually decreased to 184 tons in 1980. It has been largely discontinued as a retarder due to its toxicity (e.g., NDDA is suspected as a nitrosating agent of secondary amines many of which are carcinogenic). The current rate of consumption of vulcanization retarders is relatively small (e.g., about 3 million pounds annually [9]) of which N-(cyclohexyl thio) phthalimide accounts for the principal amount. This and other retarder compounds now used in smaller quantities have minor effects on the cure rate, or physical properties and on the degradation of the fabricated product.

e) *Antidegradants*

Antidegradants are chemical agents which are required to arrest or retard the deterioration of rubbers with aging (particularly at elevated temperatures). The deterioration of rubbers is usually a greater problem in unsaturated than saturated rubbers, which can result from a number of processes including: a) chain breakdown, b) cross-link breakdown, c) cross-link formation, d) development of polar groups and e) development of chromophoric groups in the polymer.

Additionally, the sulphide links may undergo oxidative attack usually resulting in decreased cross-link density [3]. Many of the above processes are caused by oxidation during which the unsaturated polymer itself may be oxidized via free-radical mechanisms in a chain reaction mediated by peroxide and a hydroperoxide produced during the process. This reaction is auto-catalytic and thus proceeds with increasing velocity. If two hydroperoxides react, a rubber alkoxy radical and another rubber peroxide form. Both of the radicals then remove hydrogen from polymer chains and the polymer destruction accelerates [3, 9].

The effect obtained from the above processes varies, depending on the kind of rubber polymer involved. For example, natural or polyisoprene rubbers soften as the result of chain-breaking and depolymerization. SBR and specialized rubbers such as the urethanes, fluorinated elastomers, acrylics, and some silicone elastomers harden because of cross polymerization [9].

Antioxidants which have been found effective against these processes act either by interrupting chain reactions or by preventing free-radical formation. The most common antioxidants that react with the peroxides formed by removal of hydrogen atoms are phenols, certain alkyl and aromatic amines [3, 9].

Antidegradants, which include both antioxidants and antiozanants, account for the largest volume of chemicals used in the rubber industry as additives [9].

Antioxidants, the principal type of antidegradant have a long history dating back to more than 100 years ago when fabricated rubber products were shorttlived. Prior to 1910, pitch, creosote, and naphthalene from coking coal were employed to protect rubber against deterioration [9].

In subsequent years a variety of compounds were employed including: aniline, various cresols, hydroquinone, phenol and various simple amines and hydroxy derivatives of these compounds.

The introduction after World War II of an increasing number of new elastomers and specialty elastomers increased the need for both nondiscoloring antioxidants as well as more efficient antioxidants to cover the range of synthetic elastomer structures. A large number of alkyl and aromatic amines are currently used as antioxidants resulting from the initial development of phenyl-beta naphthylamine (PBNA) the first effective long-term antioxidant commercially introduced in 1928. PBNA was used generally at levels of 1–2% in rubber processing to impart heat, oxidation and flex-cracking resistance in natural rubber, synthetic rubbers and latexes. Its use was discontinued in the late 1970's when it was dislocated that PBNA is metabolized in humans to the carcinogen 2-naphthylamine [12].

The largest volume amine antidegradants currently used are the p-phenylenediamine derivatives accounting for 50% of the approximately 166 million pounds produced in 1986 [9]. The volatility, resistance to direct oxidation, and other properties of these compounds vary widely according to the substituent groups. In addition to being powerful antioxidants, the p-phenylenediamines possess antiozonant properties. The p-phenylenediamines with both alkyl and aryl substituent groups (e.g., N,N'-p-phenylenediamine) are frequently blended with other types of antioxidants to modify either the physical form or the performance characteristics.

The phenolics constitute the second largest groups of antidegradants employed accounting for approximately 21 million pounds of antidegradants produced in the U.S. in 1986 [9]. The phenols, including bisphenols, have relatively large substituent groups and are generally known as hindered phenols. The substituent groups may include styrene, alkyls of varying complexity or combinations of these. The most widely used phenolic antioxidant is 2,6-di-tert-butyl-p-cresol. Other phenolic antioxidants are polyphenols and hydroquinone derivatives (e.g. nonobenzylether of hydroquinone). Many mixtures and blends of the phenolic antioxidants are commercially available for use in special circumstances.

The phosphites constitute an additionally important class of antidegradant whose use is projected to equal or possibly surpass the volume of phenolics consumed in 1986–1987 [9]. Both alkyl and aryl phosphates are used with tris-nonylphenol phosphite being a principal example [10].

The quinolines constitute the fourth largest group of antidegradants used constituting approximately 21 million pounds of antidegradants produced in the U.S. in 1986 [9]. Polymerized 2,2,4-trimethyl-1,2-dihydroquinoline is used most frequently where good aging and heat resistance are desired. Other commercially important amine antidegradants are the diphenylamines and various aldehyde-amine (e.g., acetaldehyde-aniline products and ketone-amine (e.g., diphenylamine-acetone product). Diphenylamine can be combined with various alkyl and aromatic groups to produce useful antioxidants. Approximately 9.7 million pounds of diphenylamines were consumed as antidegradants in 1986 [9].

Although ozone is normally present in trace amounts in the atmosphere, it is important in rubber deterioration causing deep cracks perpendicular to the direction of stress in unprotected rubbers. This reaction is initiated via the addition of ozone to the double bond of the unsaturated rubber, followed by rearrangement and hydrolysis [3].

Many of the antioxidants are also effective antiozonants. However the most important are the various p-phenylenediamines. Other antiozonants which are used in rubber processing include derivatives of urea, thiourea, and certain metallo-dithiocarbamates although their volumes consumed are relatively small compared to the p-phenylenediamines. Although the branched alkyl aryl-p-phenylenediamines are widely used for protection against ozone, the mechanism by which they act is not completely clear [3, 9].

Although organic antiozonants are used extensively, it should be noted that waxes were employed originally which act by diffusing slowly to the surface or the rubber forming a protective film (bloom) that protects the rubber surface physically from ozone attack. Waxes are still employed on a limited basis, especially in static applications. When the film is broken by flexing, however, ozone will attack at the break harming the rubber more than it possibly would otherwise [3, 9].

f) *Blowing Agents*

Blowing agents are used to produce foam rubbers by decomposing at curing temperatures to produce gases. The major blowing agents and routes of decomposition are shown in figure 4 [3]. They include: dinitrosopentamethylene-tetramine, azobisbutyronitrile, azobisformamide, benzenesulfonyl hydrazide, and p,p'-oxybis (benzenesulphonyl hydrazide). Sodium bicarbonate has also been widely used in sponge compounding, as well as with organic compounds listed above.

g) *Plasticizers*

Plasticizers are generally employed to reduce the viscosity of rubber in order to process and incorporate fillers and compounding ingredients. A large variety of plasticizers are employed at levels generally not exceeding 30 percent by weight of rubber product. The major classes of agents used in this capacity include: phthalate esters (dioctyl-, diisodecyl, -didecyl-,); adipate esters (dibenzyl-, didecyl-, diisodecyl, dioctyl-); sebacate esters (dibutyl-, dioctyl-, disoctyl); cumarone-indene resins, triethylene glycol dicaprylate, triethylene glycoldicaprylate, triethylene glycoldicaprylate, zinc-2-benzamido thiophenate and many fatty acids and terpene resins.

h) *Processing Aids*

A large variety of agents are employed as processing aids primarily to make uncured rubber softer and more readily mixed, extended or calendered. Although pine tars were used initially, these have been almost completely replaced by paraffinic, naphthenic and aromatic mineral oils with the latter more extensively employed. The aromatic oils are principally obtained from solvent-refining lubricating and

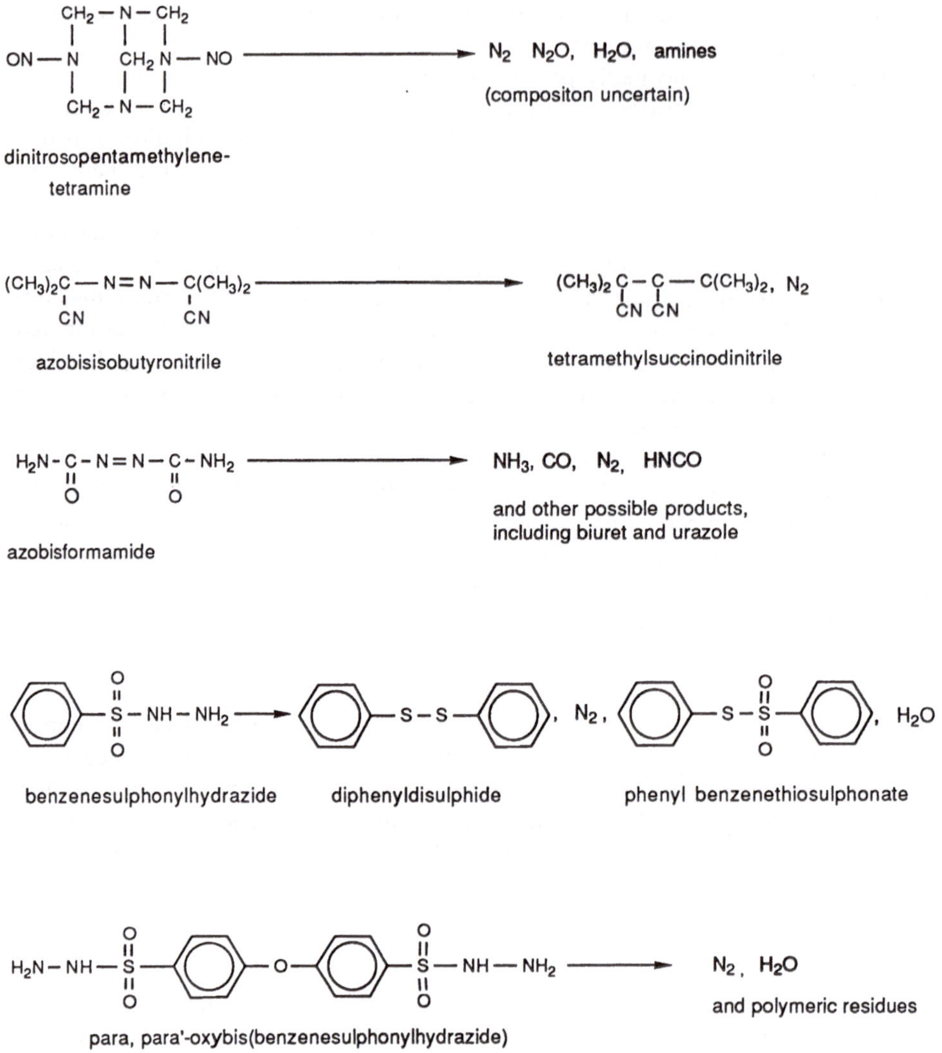

CH₂ — N — CH₂
| | |
ON — N   CH₂ N — NO ──────────────→ N₂ N₂O, H₂O, amines
| | |
CH₂ - N — CH₂   (compositon uncertain)

dinitrosopentamethylene-
      tetramine

(CH₃)₂C — N=N — C(CH₃)₂ ──────────────→ (CH₃)₂ C — C — C(CH₃)₂, N₂
| | | |
CN   CN   CN CN

     azobisisobutyronitrile   tetramethylsuccinodinitrile

H₂N - C - N = N — C - NH₂ ──────────────→ NH₃, CO, N₂, HNCO
     || ||
     O   O   and other possible products,
                                                  including biuret and urazole
azobisformamide

benzenesulphonylhydrazide ──→ diphenyldisulphide , N₂ , phenyl benzenethiosulphonate , H₂O

H₂N — NH — S(...)— O —(...)— S — NH — NH₂ ──────────→ N₂, H₂O
                                                   and polymeric residues

para, para'-oxybis(benzenesulphonylhydrazide)

Fig. 4. Some blowing agents and their routes of decomposition

cutting oils and can contain relatively large quantities of polycyclic aromatic hydrocarbons [3]. For example, 4–6 ring polycyclic hydrocarbons have been reported to be
typically present at a concentration of approximately 20% in these oils and a typical
aromatic processing oil may contain 50 ppm of benzo (a) pyrene [3, 13]. Other
materials which have been used as processing aids include: talc, coal-tar pitch,
vegetable oils, phthalates, organic phosphates and polymerisates of unsaturated
vegetable or animal oils with sulphur or sulphur chloride [3].

Additionally, peptizing agents such as benzimidazoles, thiophenols and mer-
captobenzothiazole are occasionally used to soften the rubber before chemical interaction. These agents contain thiol groups which function as chain terminating agents
for the radicals formed by rupture of the polymer chains during mastication.

i) *Reinforcing Agents, Fillers and Diluents*

Many rubbers are compounded with reinforcing agents and fillers (with levels in some cases as high as 10 to 15% in finished products) to improve their tensile strength and abrasion resistance and reduce costs. Of the reinforcing agents, the most important agents are carbon black and amorphous silica. The original form of carbon black was lamp black. However, during the period 1910–1942 lamp black was largely superseded by channel black which was produced from natural gas by burning in small sooty flames impinging on a coal surface (e.g., channel iron). The channel blacks have been almost completely replaced since the 1940's by furnace blacks. Although the furnace blacks were originally made from natural gas by the furnace process, they are now produced from oil. Furnace blacks contain a higher level of polycyclic aromatic hydrocarbons than channel blacks [3]. The U.S. Food and Drug Administration restricts the carbon black content of packaging materials used in contact with foodstuffs. For example, the furnace combustion black content is not to exceed 10% by weight of rubber products intended for use in contact with milk or edible oils [14].

Two types of synthetically produced silica are used as rubber reinforcement agents. These are: a) precipitated silica made by direct precipitation from sodium silicate solution and b) pyrogenic silica made by reacting silicon tetrachloride with water vapor in a hydrogen-oxygen flame [3].

A number of fillers and diluents are employed for non-critical applications without detracting from the properties of the vulcanisate. These include: calcium carbonate, clays, barytes, magnesium carbonate, barium sulfate, aluminum silicate, zinc carbonate, zinc sulfide and titanium dioxide [3, 14]. The clays may contain crystalline silica, e.g., "China" clays may contain up to 1% and "ball" clays up to 20% silican [3].

j) *Pigments*

A large number of both inorganic pigments and organic dyestuffs are used to produce coloured rubbers. Carbon blacks are among the most important rubber pigments. Titanium whites are most widely used for imparting whiteness. They are either pure titanium dioxide or coprecipitates of titanium dioxide with insoluble barium or calcium salts. Other inorganic pigments that are used include: chrome oxides (Cr_2O_3), iron oxide, zinc chromate and calcium carbonate. The organic dyes include: C.I. pigment red, phthalocyanine, phthalocyanine blue, and ultramarine blue [3, 14, 15]. The above organic dyes as well as the inorganic pigments chrome oxide, iron oxide, titanium dioxide and zinc chromate are limited by the U.S. Food and Drug Administration [14] to amounts not to exceed 10% by weight of rubber product in rubber articles intended for repeated use.

k) *Bonding Agents*

Bonding agents are used to bond rubber to the textile used in the construction of tires or rubber/metal products. Although many of the systems employed are proprietary mixtures often of undisclosed composition, it is believed that they may contain isocyanates and/or p-dinitrosobenzene. Additionally, resorcinol-hexamethylene tetramine bonding systems are also used [3].

l) *Solvents*

A large variety of organic solvents are used in compounding and processing of rubber. These include: aliphatic hydrocarbons, acetone, methyl ethyl ketone, methylene chloride, perchlorethylene, trichloroethylene, 1,1,1-trichloroethane, benzene, toluene, xylene, tetrahydrofuran and dimethylformamide [3]. As will be noted later a number of these solvents have been reported to be carcinogenic and/or genotoxic.

m) *Miscellaneous Rubber Chemical Additives*

(i)   *Mould-release Agents* include soaps, synthetic detergents, silicones, fluorinated hydrocarbons and polyethylene which are generally sprayed on to moulds to facilitate the removal of the cured rubber items [3].

(ii) *Antitack Agents.* The principal antitack agents employed are mica, talc and metal stearates (e.g., zinc stearate) [3].

(iii) *Emulsifers.* The major emulsifiers employed in the rubber industry include: rosins and rosin-derivatives, sodium decylbenzenesulfonate, sodium dodecyl-benzenesulfonate, sodium lauryl sulfate, tall oil mixed soap (calcium, potassium and sodium), naphthalene sulfonic acid-formaldehyde condensate (sodium salt) and sodium or potassium fatty acid salts [14].

(iv) *Miscellaneous Rubber Additives.* Additional categories of rubber additives include *flame-proofing agents*, (e.g., antimony trioxide, halogenated aliphatic hydrocarbons, aluminium hydrates); *coupling agents* such as silanes and *odorants*, such as blends of perfumes and oils and freezing point depressants and elasticizers [3].

(v)   *Latex Processing Chemicals.* Materials which are unique to latex processing are *dispensing agents*, such as casein and a sodium alkyl sulphate; *wetting agents*, such as alkyl sulphonates, organic phosphates or a nonyl phenol-ethylene oxide adduct; *thickeners*, such as polyacrylates, cellulose derivatives or other high-molecular weight carbohydrate derivatives; *preservatives*, such as formaldehyde and pentachlorophenates; *coagulants*, such as sodium silicofluoride and *foaming* and *antifoaming* agents [3].

(vi) *Reclaiming Agents.* Vulcanized rubber scrap is converted to plastic processable material by a reclaiming process. Reclaiming is essentially the depolymerization of vulcanized rubber which involves heating waste vulcanized rubber with reclaiming agents which include: di- and trialkylphenol sulfides and disulfides, thiocresol, dibutyl-m-cresol, aryldisulfides and calcium and zinc chlorides [14].

n) *Nitroso Compounds*

It is important to note the occurrence of nitrosation resulting in the formation of volatile N-nitrosamines during the manufacture and processing of rubber [1, 3, 16–22]. The following N-nitrosamines have been detected which are among the most potent carcinogens known [21]: N-nitrosodimethylamine, N-nitrosodiethylamine, N-nitrosodibutylamine, N-nitrosopiperidine, N-nitrosopyrrolidine and N-nitrosomorpholine. It should be stressed that the corresponding nitrosatable precursor of each of these compounds (with the exception of N-nitrosopyrrolidine) is used in rubber compounding [3].

In addition, to the well recognized nitrosation of secondary amines, reactions involving tertiary amines or amine derivatives (e.g., dialkyl dithiocarbamates, dialkyl thiuram sulfides and sulphenamines) are possible since these types of agents are used as vulcanization accelerators and stabilizers. Figure 5 illustrates the groups of chemicals that can be converted to nitrosamines. These include: secondary and tertiary amines, quarternary ammonium salts, dithiocarbamates, sulfenamides, thiuram sulfides and N-disulfide derivatives [3]. Nitrosation can occur in aqueous and solid systems as well as in the gas phase.

Accelerators and stabilizers derived from dialkylamines are frequently employed in elastomer compounding. Polymers compounded with dialkyl amine accelerators or stabilizers have been found to contain the corresponding extractable N-nitrosamines. For example, dipentamethylene thiuram disulfide yielded N-nitrosopiperidine and zinc dibutyl-dithiocarbamate yielded N-nitroso-di-n-butylamine. When two different compounds were present such as in chlorosulfonated polyethylene and ethylene-propylenediene terpolymer, then both the corresponding N-nitrosamines were detected [18]. The elastomer need not be cured for N-nitrosamines to be present. If the raw polymer contains a dialkylamino compound, then the corresponding N-nitrosamine can be extracted.

Previous studies have shown that N-nitrosamines are evolved from heating elastomers [22]. Additionally, Ireland et al. [18] demonstrated that N-nitrosamines could be extracted in water from elastomers. The level of nitrosamines extracted ranged from 0% to 1,400 ppb. from compounded and cured polychloroprene, ethylene-propylene-diene terpolymer, chlorosulfonated polyethylene, and natural rubber as well as from several commercial rubber articles. This study did not identify a nitrosating agent that could act upon the dialkylamino compounds during processing [18].

A number of accelerators have been shown to yield volatile nitrosamines following nitrosation as illustrated in figure 6. These include: zinc dibenzyl dithio-

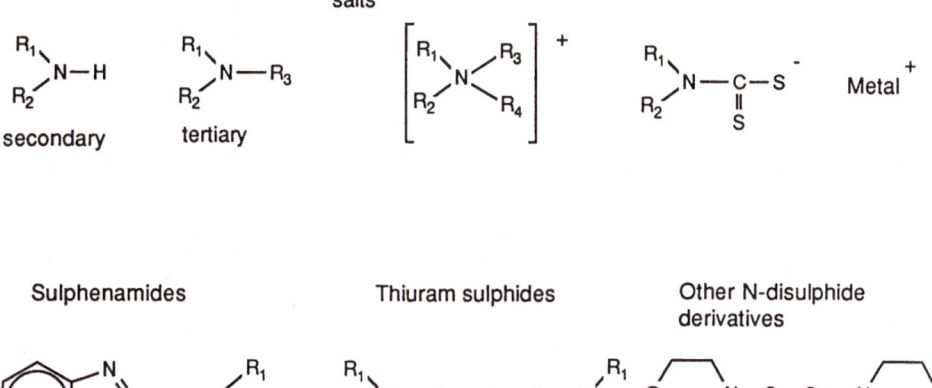

Fig. 5. Groups of chemicals that can be converted to *n*-nitrosamines

carbamate, N,N'-dicyclohexyl-2-benzothiazolesulfenamide, zinc ethylphenyldithiocarbamate and dimethyl diphenylthiuram disulfide [3].

A number of additional chemicals used in the rubber industry as additives contain
nitro or nitroso groups that are potential nitrosating agents. Since some of these agents
may decompose thermally during processing to yield nitrogen oxides, direct nitrosation of nitrosatable compounds or precursors in solid rubber must be considered
possible [3, 23].

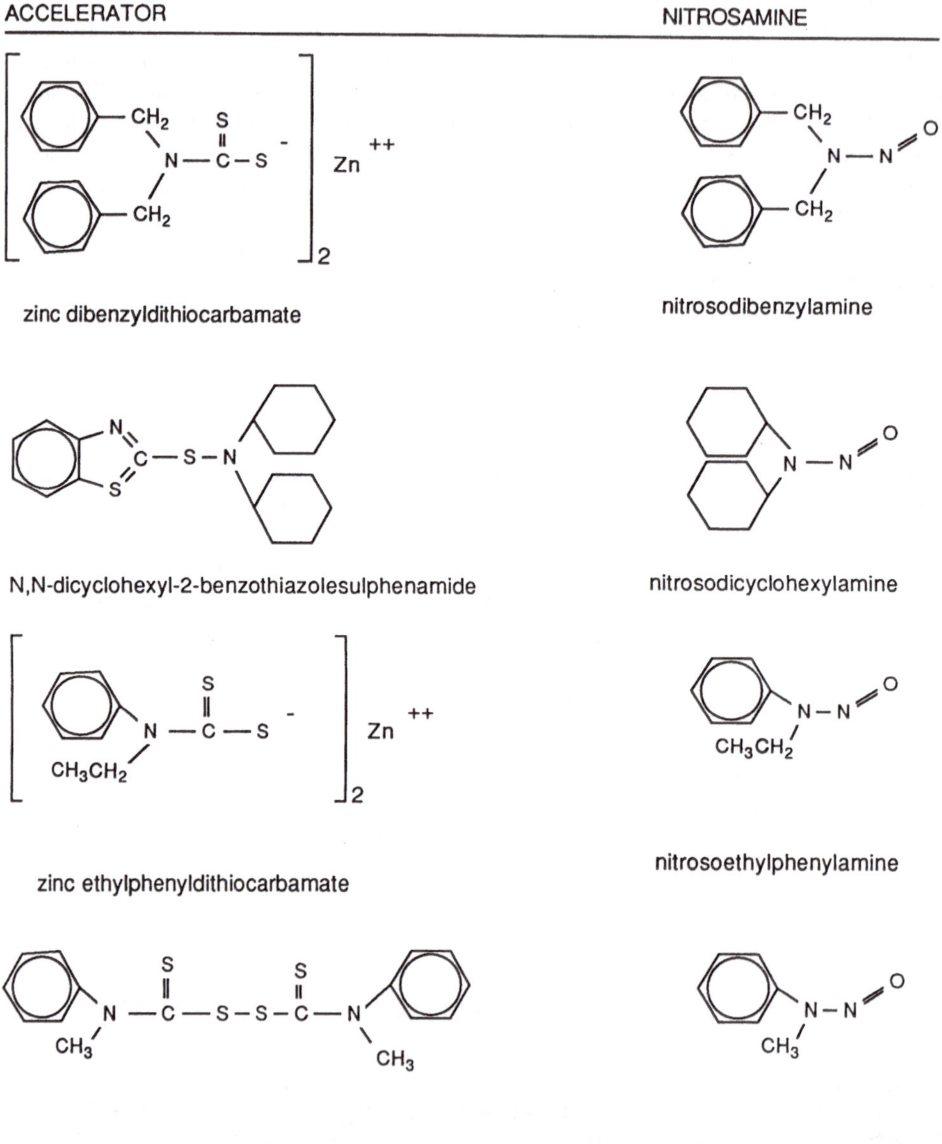

ACCELERATOR   NITROSAMINE

zinc dibenzyldithiocarbamate   nitrosodibenzylamine

N,N-dicyclohexyl-2-benzothiazolesulphenamide   nitrosodicyclohexylamine

zinc ethylphenyldithiocarbamate   nitrosoethylphenylamine

dimethyldiphenylthiuram disulphide   nitrosomethylphenylamine

Fig. 6. Accelerators and corresponding non-volatile nitrosamines

A number of nitro and nitroso compounds that are used in the rubber industry which can act as precursors of nitrogen oxides are shown in Fig. 7 [3]. These include the retarders N-nitrosodiphenylamine and poly-N-nitroso-2,2,4-trimethyl-1,2-dihydroquinoline; the accelerators 2-(2,4-dinitrophenylthio)benzothiazole and N,N-dimethyl-p-nitrosoaniline; the promoter N-(2-methyl-2-nitropropyl)-4-nitroso-aniline and the blowing agent dinitrosopentamethylenetetramine.

It should also be restressed that nitrosamines can also be present in rubber chemicals as contaminants. The origin of these nitrosamines are believed to result as a consequence of nitrosation via ambient nitrogen oxides during production or storage [21]. Examples of nitrosamine contamination in seven commercial samples of accelerators are illustrated in Table 3 [3, 21].

RETARDERS

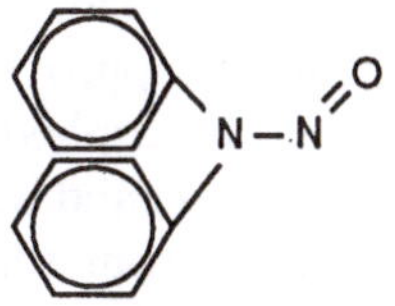

N-nitrosodiphenylamine

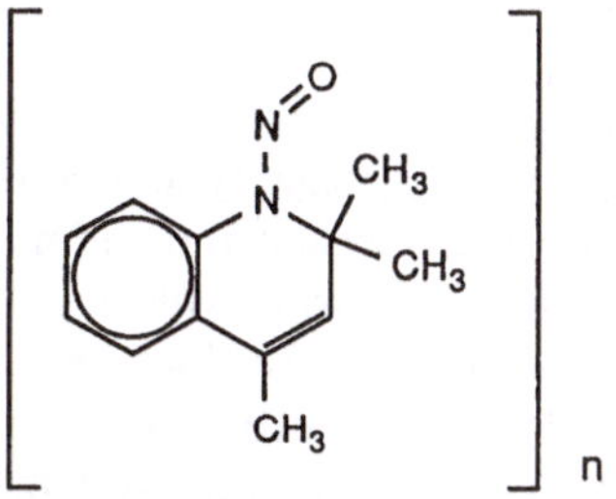

poly-N-nitroso-2,2,4-trimethyl-
1,2-dihydroquinoline

ACCELERATORS

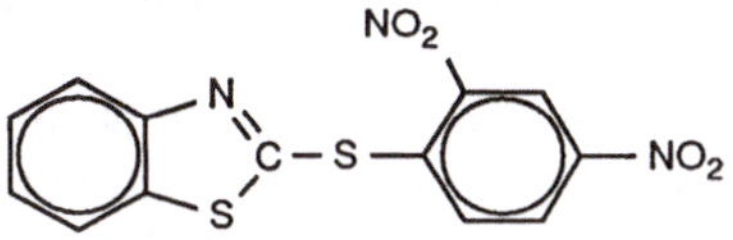

2-(2,4-dinitrophenylthio)-
benzothiazole

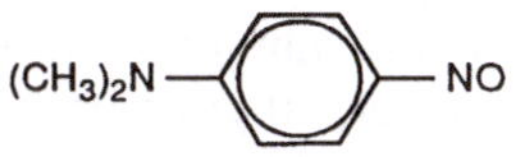

N,N-dimethyl-para-nitrosoaniline

PROMOTOR

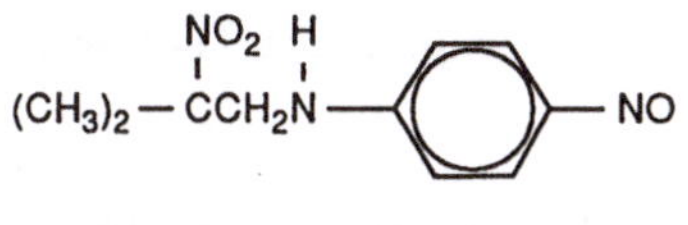

BLOWING AGENT

N-(2-methyl-2-nitropropyl)-
4-nitrosoaniline

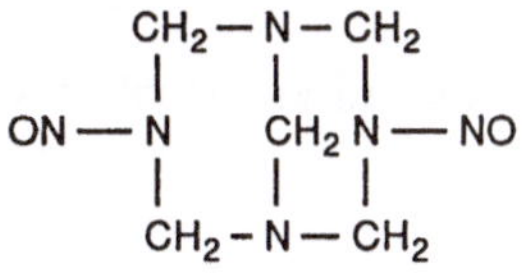

dinitrosopentamethylenetetramine

Fig. 7. Nitro and nitroso compounds used in the rubber industry that can act as precursors of nitrogen oxides

Table 3. Examples of nitrosamine contamination in commercial samples of rubber chemicals

Accelerator	Nitrosamine present	Concentration (µg/kg)
N-pentamethylene dithiocarbamate piperidine salt	N-nitrosopiperidine	200
Tetramethylthiuram disulfide	N-nitrosodimethylamine	4–800
Tetraethylthiuram disulfide	N-nitrosodiethylamine	25–80
Zinc pentamethylene dithiocarbamate	N-nitrosopiperidine	60–450
Zinc dibutyldithiocarbamate	N-nitrosodibutylamine	65–2500
Zinc diethyldithiocarbamate	N-nitrosodiethylamine	10–100
Morpholine derivatives	N-nitrosomorpholine	60–3500

Chemical By-Products

With the exception of N-nitrosamines there are very few data relating to the nature of the chemical species generated during the pyrolysis of rubber. Pyrolysis reactions do not normally occur and the maximum temperature of vulcanization is approximately 240 °C. During vulcanization the mould contains a reducing atmosphere since air is excluded. Products that could thus be formed during vulcanization and be released from the surface include amines and organic sulfides derived from the accelerators (e.g., curing systems containing tetramethylthiuram disulfide and dimethylamine during curing) [3, 24]. The accelerator N-*tert.*-butyl-2-benzo-thiazole sulfenamide can yield tert. butylisothiocyanate at curing temperatures [3, 25]. Additionally, sulfenamide accelerators have been reported to yield mercaptobenzo-thiazole and characteristic amines [3, 26–29], while thiourea accelerators decompose to yield isothiocyanates (e.g., diethylthiourea produces ethyl isothiocyanate) [3, 30]. The retarding agent cyclohexylthiophthalimide was been shown to be converted to phthalimide [31].

A study of the mutagenicity of rubber vulcanization gases derived by heating chloroprene rubber, ethylene-propylene rubber, styrene-butadiene rubber and buta-diene-acrylonitrile rubber indicated a very complex situation with respect to the number of and properties of substances responsible for the mutagenicity (curing systems, additives and filler materials from various sources were represented in the material) [2, 32].

Major Monomers and Rubber Additives of Carcinogenic, Genotoxic and Teratogenic Concern

The number and diversity of chemicals used individually and/or in admixture as we have noted is very large indeed. The major biological effects of concern are carcinogenic, genotoxic and teratogenic and these will be noted for the principal agents where reported in animals (experimental systems) and to a much more limited extent in humans.

a) *Monomers*

The monomers used in the rubber industry are listed in Table 4 [3]. Small amounts of monomers, e.g., acrylonitrile, butadiene, chloroprene, ethylene, propylene styrene, vinylacetate, vinyl chloride and vinylidene chloride may remain in solid rubber and could be released into the work environment [3]. The major monomers employed in the rubber industry that are of toxicological concern are butadiene, acrylonitrile, styrene, vinyl chloride, vinylidene chloride, chloroprene and ethylene oxide.

Table 4. Monomers used in the rubber industry

Monomers

Acrylonitrile
1,3-Butadiene
Chloroprene (2-chloro-1,3-butadiene)
Chlorotrifluoroethylene
Dicyclopentadiene
Epichlorohydrin
Ethyl acrylate
Ethylene
Ethylene oxide
Ethylidene norbornen
1,4-Hexadiene
Hexafluoropropylene
Hexafluoropropylene/tetrafluoroethylene
Isobutylene
Isoprene (2-methyl-1,3-butadiene)
Methylene diphenyldiisocyanate[a]
Methylmethacrylate
Naphthalene-1,5-diisocyanate[a]
Polymethylene polyphenyl isocyanate[a]
Propylene
Propylene oxide
Styrene
Tetrafluoroethylene[a]
2,4-Toluenediisocyanate[a]
2,6-Toluenediisocyanate[a]
Vinyl acetate
Vinyl chloride
Vinylidene chloride

[a] Not technically monomers but used to make some elastomers

(i) *1,3-Butadiene*

Butadiene, as the 1,3-isomer, is consumed in far larger quantities than any other monomer for synthetic rubber. In the U.S., more than 20% of butadiene consumption (2.75 billion pounds in 1987) goes into elastomers principally styrene-butadiene rubber, polybutadiene, polychloroprene and nitrile rubbers [6].

1,3-Butadiene was found carcinogenic when tested by inhalation in $B6C3F_1$, mice of both sexes producing an unusual neoplasm of the heart (hemangiosarcoma) as well as tumours at several other sites including lung, stomach, liver, mammary gland and the lymphatic system in a dose-related manner [33, 34]. 1,3-Butadiene was also associated with non-neoplastic lesions of the respiratory epithelium, liver necrosis and testicular atrophy [33].

1,3-Butadiene has been found mutagenic using the Ames bioassay employing *S. typhimurium* tester strain TA 1530 (a base pair substitution strain) in the presence of an S9 metabolic system [34, 35]. It was not mutagenic in this strain in plate incorporation and liquid preincubation assays [35, 36]. 1,3-Butadiene is not a direct-acting mutagen and it has been suggested that epoxides such as 1,2,3,4-diepoxy-butene, 1,2-epoxybutene, and/or 3,4-epoxy-1,2-butane diol are probably the ultimate mutagenic forms of 1,3-butadiene [33, 34, 37].

In vivo sister chromatid exchange (SCE) and micronucleus studies with 1,3-buta-diene administered by inhalation to B6C3F, mice and Sprague-Dawley rats showed a significant dose-dependent increase in both micronuclei (MN) and SCE's in mouse bone marrow cells. In contrast, rat bone marrow cells did not exhibit increases in MN or SCE induction [38]. This paralleled the chronic bioassays which showed mice to be more susceptible than rats to 1,3-butadiene carcinogenicity [33, 34].

Butadiene monoxide, a presumed metabolite of 1,3-butadiene was shown to be a very effective inducer of SCE's and chromosome aberations after *in vivo* exposure to mice [39].

Although there is no information on the metabolism of 1,3-butadiene in humans, a scientific basis for the extrapolation from animal metabolism data to humans for 1,3-butadiene is provided by the similarities in the epoxidation of the isolated double band in benzo(a)pyrene by organ and tissue cultures from animal and human sources [40].

IARC has evaluated the available case-report studies and noted that they provided some evidence of an association between an excess of leukemia and lymphomas and work in an environment with mixed exposure to 1,3-butadiene and other chemicals, especially styrene and possibly benzene. The mixed exposure patterns typical of the rubber industry however rendered it impossible to single out 1,3-buta-diene as a specific causative agent. Hence IARC stated that "there is *inadequate* evidence for the carcinogenicity of 1,3-buradiene in humans" while there is *sufficient* evidence for the carcinogenicity of 1,3-butadiene in experimental animals [34].

(ii) *Acrylonitrile*
Acrylonitrile, used in nitrile elastomers as a copolymer with butadiene is produced at a rate of about 2.5 billion pounds per year in the U.S. Nitrile elastomers currently require about 50 million pounds of acrylonitrile in the U.S. [6].

Acrylonitrile induced neoplasms of the brain, squamous cell papillomas of the stomach and Zymbal-gland carcinomas as well as tumors of the small in-testine and mammary gland in the rat following its oral administration [41–43]. Following its inhalation, neoplasms of the central nervous system, mammary glands, Zymbal glands and forestocham were observed in rats [41, 44]. IARC [45] considers the evidence for carcinogenicity of acrylonitrile to be *sufficient* to animals and *limited* to humans (IARC Group 2A categorizes acrylonitrile as *probably* carcinogenic to humans) [45].

The mutagenicity of acrylonitrile has been demonstrated in the *Salmonella/*

microsome assay system with metabolic activation [46–48] and in some strains of
E. coli (without metabolic activation) [49]. Results of in vivo studies with acrylonitrile are contradictory. It did not induce chromosome aberrations in bone marrow
cells of micronuclei in polychromatic erythrocytes of mice and rats [39]. A dominant
lethal test to detect chromosome aberrations in male mice germ cells also yielded
negative results [50–52]. However, Kawachi et al. [73] obtained positive results in rat
bone marrow cells *in vivo*.

Teratogenic effects of acrylonitrile have been reported in the offspring of rats
treated during pregnancy either by gavage [54], by inhalation [55] or in the drinking
water [56].

(iii) *Styrene*
Styrene, consumed mainly in styrene-butadiene rubber (SBR), various styrene-
butadiene latices and thermoplastic elastomers, ranks next to butadiene as the
largest-volume monomer going into elastomers. In the U.S. in 1987, those uses
totaled more than 800 million pounds. SBR however required only about 40%
of the total going into elastomers, a little more then 10% of total production of about
7.9 billion pounds [6].

Oral administration of styrene to mice and rats, resulted in an increased incidence
of lung tumours only in male mice [57]. IARC [45] considers the evidence for
carcinogenicity to animals for styrene to be *limited*. IARC [45] has evaluated the
evidence of carcinogenicity for styrene to humans to be *inadequate*.

Styrene is metabolized in humans and mammals to styrene oxide. There is
sufficient evidence for the carcinogenicity in experimental animals of styrene oxide
[58].

In animals treated *in vivo*, styrene induced micronuclei, SCE's and DNA strand
breaks, with conflicting results obtained for chromosane aberrations [45]. In
human lymphocytes *in vitro*, styrene induced chromosomal aberrations, micronuclei
and SCE's [45]. Styrene oxide was mutagenic to *S. typhimurium* TA 1535 and
TA 100 in the absence of metabolic activation [41, 59, 60].

Inhalation gestational exposures of styrene oxide to rats and rabbits produced
reproductive and development toxicity as well as maternal toxicity [61].

(iv) *Chloroprene (2-Chloro-1,3-butadiene)*
Chloroprene has been utilized for the production of neoprene elastomers. The production of chloroprene is estimated to total about 250 million pounds annually [6].

The evidence of carcinogenicity of chloroprene to both animals and humans has
been judged by IARC to be *inadequate* [45].

An increased incidence of chromosomal aberrations was found in the lymphocytes of workers exposed to chloroprene [45]. Chloroprene has been shown to induce
dominant lethal mutations in rats and chromosomal aberrations in bone-marrow
cells of mice treated *in vivo*. Although it induced transformation in one hamster cell
line, it did not induce mutations in Chinese hamster cells. Chloroprene induced sex-
linked recessive lethal mutations in *Drosophila* and was mutagenic to bacteria
[45].

(v) *Vinyl Chloride (VCM)*
Vinyl chloride monomer (VCM) has been associated in humans with a broad range
of tumours including the brain, liver, lung and haematolymphopoietic system [41].

A large number of epidemiological studies [41, 45] and case reports [41, 45] have substantiated the causal association between vinylchloride and angiosarcoma of the liver. Additional studies also indicate that exposure to vinyl chloride causes other forms of cancer, e.g., hepatocellular carcinogenicity, brain tumours, lung tumours and malignancies of the lymphatic and haematopoitic system [41, 45]. The evidence for carcinogenicity in humans is thus considered by IARC to be *sufficient* [45].

Additionally, vinyl chloride when administered orally or by inhalation to mice, rats and hamsters produced tumours in the mammary gland, lung, Zymbal gland and skin and angiosarcoma of the liver [41, 45] and hence the evidence for carcinogenicity to animals is considered *sufficient* by IARC [45].

Chromosomal aberrations were induced in peripheral blood lymphocytes of workers exposed to VCM at levels of 5–500 ppm (13–1300 mg/m^3) [45]. VCM induced chromosomal aberrations, SCE's and micronuclei in rodents exposed *in vivo* but did not induce mutation in the mouse spot test or dominant lethal mutations in rats or mice. It induced mutations in Chinese hamster cells and unscheduled DNA synthesis in rat hepatocytes *in vitro* and induced transformation of BALB/C3T3 cells [45].

(vi) *Vinylidene Chloride (1,1-Dichloroethylene)*

Vinylidene chloride when administered by inhalation to mice induced an increase in the incidence of kidney adenocarcinomas in male mice as well as an increase in mammary carcinomas in females and of pulmonary adenomas in male and female mice.

In inhalation studies, no treatment-related neoplasm was observed in rats or hamsters. Oral administration of vinylidene chloride to mice and rats yielded negative results [34]. The evidence for carcinogenicity of vinylidene chloride to animals was considered by IARC to be *limited* and the evidence for carcinogenicity in humans to be *inadequate* [45].

Vinylidene chloride is mutagenic to bacteria, plant cells as well as induces mutation and gene conversion in yeast. Although it induced unscheduled DNA synthesis in treated mice, it did not induce dominant lethal mutations in mice or rats and did not induce chromosomal aberrations in bone-marrow cells of rats treated *in vivo* [34, 45].

(vii) *Ethylene Oxide*

The evidence for carcinogenicity of ethylene oxide in humans is considered by IARC to be *limited* [45]. Although a causal relationship between exposure to ethylene oxide and leukemia was considered possible, the 5 small epidemiological studies thus far available are believed to suffer from various disadvantages, principally confounding exposures making their interpretation exceedingly difficult [45, 58].

The evidence for carcinogenicity of ethylene oxide to animals is considered by IARC to be *sufficient* [45, 58], based on inhalation studies in mice and rats. In mice, inhalation of ethylene oxide resulted in increased incidence of alveolar/bronchiolar lung tumours and tumours of the Harderian gland in animals of each sex and of uterine adenocarcinomas, mammary [45, 62] carcinomas and malignant lymphomas in females.

Rats exposed by inhalation showed increased incidences of mononuclear-cell leukemia, brain tumours and proliferative lesions of the adrenal cortex in animals of each sex and of peritoneal mesotheliomas in males [45, 58, 63].

Ethylene oxide induced chromosomal aberrations and SCE's in peripheral lymphocytes of monkeys exposed *in vivo*. Rodents exposed to ethylene oxide resulted in induced chromosomal aberrations, SCE's, micronuclei, dominant lethal mutations, heritable translocation, and in the alkylation of hemoglobin and DNA. In human cells *in vitro*, ethylene oxide induced SCE's, chromosomal aberrations and unscheduled DNA synthesis. Ethylene oxide induced somatic and sex linked recessivé lethal mutations and heritable translocations in *Drosphila*. It was mutagenic in bacteria and fungi [45, 64].

b) *Accelerators, Antidegradants and Curing Agents*

2-Mercaptobenzothiozole (2-benzothiazolyl mercaptan) when administered to F344/N rats by gavage for 2 years, resulted in increased incidences of mononuclear cell leukemia, pancreatic acinar cell adenomas, adrenal gland pheochromocytomas and preputial gland adenomas or carcinomas (combined) in male rats. In female F344/N rats, incidences of adrenal gland pheochromocytomas and pituitary gland adenomas [65] were reported.

There was no evidence of carcinogenic activity of 2-mercaptobenzothiazole for male B6C3F$_1$ mice and *equivocal* evidence of carcinogenic activity for female B6C3F$_1$, mice as indicated by increased incidences of hepatocellular adenomas or carcinomas (combined) [65].

Additional major accelerators and curing agents which are commonly used are thiuram compounds including: tetramethylthiuram disulphide (thiram, TMTD), tetraethylthiuram disulphide (TETD, disulfiram and tetramethylthiuram monosulphide (TMTM). Carcinogenicity studies with TMTD and TETD were considered either insufficient or too limited to evaluate although in one study of oral administration of TETD to mice, the incidence of liver-cell tumours was increased in males of two strains and the number of lung tumours was increased in males of oñe strain [3, 66]. It should also be stressed that both TMTD and TETD react with nitrite to form carcinogenic nitroso dialkylamines [3]

Mutagenicity studies on rubber accelerators revealed that TMTD and TMTM were mutagenic to *S. typhimurium* while TETD was inactive. Five of the nine tested dithiocarbamates were also mutagenic with ziram the most active [67].

An additional mutagenicity study involving the *Salmonella*/microsome assay found 4,4'-dithiomorpholine, zinc ethyl phenyl dithiocarbamate, N,N'-dicyclohexyl-p-phenylenediamine, N,N'-diphenyl-p-phenylenediamine, 4,4'-diamino-diphenylmethane, 6-ethoxy-1,2-dihydro-2,2,4-trimethylquinoline and N-methyl-N,4-dinitroso-aniline to be all direct or indirect mutagens [3, 68].

TMTD, TMTM and TETD were also reported to produce deaths and malformations in chicken embryos [3, 69]. Cadmium and zinc diethyldithiocarbamate, zinc ethylphenyl dithiocarbamate, and zinc dibutyldithiocarbamate were found to be particularly embryotoxic when tested in chick embryos [3, 70].

Ethylene thiourea (ETU) produced high incidences of follicular carcinomas of the thyroid in rats of both sexes after its oral administration [3, 71]. The evidence of carcinogenicity of ETU to animals was judged to be *sufficient* by IARC [45]. IARC judged the evidence of carcinogenicity of ETU to humans to be inadequate [45].

ETU did not induce dominant lethal mutations, micronuclei or SCE's in mice or chromosomal aberrations in rats treated *in vivo* or in rodent cells *in vitro*. It did not induce sex-linked recessive lethal mutations in *Drosophila*, but induced aneuploidy and mutation in yeast [45, 64].

When ETU was administered perorally to rats and rabbits, it was found to be tetratogenic [72, 73]. It was a weak teratogen when tested in chick embryo, while tetramethylthiourea and 1,3-diphenylthiourea were judged to be the most active teratogens [3, 74].

1,3-Diphenylguanidine (DPG) is used primarily as a secondary accelerator to thiazoles, sulfenamides and thiurams. It has been found mutagenic in *S. typhimurium* strains TA 98 and TA 100 with hamster liver activation, while it was non-mutagenic in TA 1535 and TA 1537 with and without hamster and rat liver activation [75]. DPG was also reported mutagenic in strains TA 98, TA 100, TA 1535, TA 1537 and TA 1538 with and without metabolic activation [76].

Phenyl-2-naphthylamine (PBNA) had earlier been employed as an antioxidant for SBR often at levels of 1–2 % [12]. It has been reported that commercial samples of PBNA previously produced in the U.S. were contaminated with 20–30 mg/kg of the carcinogen 2-naphthylamine [77]. While PBNA samples in the United Kingdom in the past contained 15–50 mg/kg of 2-naphthylamine, levels of less than 1 mg/kg have been reported in recent years in one commerical product.

PBNA is a suspect carcinogen because of its metabolism to 2-naphthylamine which has been found in the urine of workers who had been exposed to PBNA and in the urine of healthy human volunteers to whom PBNA was administered [78]. A statistically significant increase in the incidence of all tumours, particularly hepatomas in male mice has been found in animals administered PBNA orally [79].

Naphthylamine-acetaldehyde condensates (NONOX-S) were introduced as antioxidants in the rubber product manufacturing industry around 1928. Their use was discontinued in the U.K. in 1949 since early products contained up to 2.5 % 1-naphthylamine and up to 0.25 % 2-naphthylamine. Exposure to Nomox S in rubber factories has been associated with an increase in the incidence of bladder cancer in workers [3, 80].

Although peroxide initiators are widely employed in the rubber industry, there is a paucity of data relating to their carcinogenic potential. Benzoyl peroxide and lauroyl peroxide have been known to be effective skin tumour promotors [81].

A number of organic peroxides have been reported to be mutagenic in earlier studies. For example cumeme hydroperoxide was mutagenic in *E. Coli* and *Neurospora*; di-tert. butyl peroxide was mutagenic in *Neurospora* and *tert.* butylhydroperoxide was mutagenic in *Drosophila E. Coli* and *Neurospora* [82].

c) *Plasticizers*

Di(2-ethylhexyl)-phthalate (DEHP), the most widely employed plasticizer has been the most extensively investigated for its toxicological properties of all the phthalic acid esters used in the rubber industry [3]. DEHP significantly increases the incidence of benign and malignant liver-cell tumours in B6C3F$_1$ mice and Fischer 344/N rats of both sexes [3, 83]. DEHP inhibits mitochondrial function and causes peroxisome proliferation, hepatomegaly and testicular atrophy in rodents. Various mutagenicity tests in bacteria and mammalian cells with DEHP showed negative or contradictory results [45].

Di(2-ethylhexyl) adipate (DEHA) is the most important adipate ester employed as a plasticizer. DEHA produced hepatocellular tumours in B6C3F$_1$ mice of both sexes when administered in a chronic feeding study [84].

d) Solvents

A large number of solvents have been or are currently used in the rubber industry [3]. Those that are of major toxicological concern include: benzene, trichloroethylene, 1,1,1-trichloroethane, methylene chloride, dimethylformamide, methyl ethylketone, and 1,4-dioxane (used as a stabilizer in solvents).

Benzene occupies a preeminent role in environmental health in regard to its production, use, occurrence, dispersion and human toxicity. Benzene was used in the rubber industry as a general purpose solvent in the 1920's but following reports of toxicity, benzene was largely supplanted by toluene and other aromatic and aliphatic solvents by the early 1950's. While benzene exposure still occurs in the industry, it is present primarily as a contaminant of other substances (e.g., toluene) and exposures are much lower currently than in the 1920's. Numerous case reports have suggested a relationship between exposure to benzene and the occurrence of various types of leukemia [45, 85]. The evidence for the carcinogenicity of benzence to humans is considered by IARC to be *sufficient* [45, 85].

Benzene is carcinogenic in mice and rats by several routes of administration. Following its oral administration at several dose levels, it induced neoplasms at multiple sites in males and females of both species [45, 85]. Exposure of rats by inhalation increased the incidence of neoplasms, mainly carcinomas at various sites [45, 85]. The evidence for carcinogenicity to animals is considered by IARC to be *sufficient* [45].

Chromosomal aberrations in human peripheral lymphocytes have also been associated with occupational exposure to benzene. It induced chromosomal aberrations, micronuclei and SCE's in bone-marrow cells of mice, chromosomal aberrations in bone-marrow cells of rats and Chinese hamsters *in vivo* [45, 64]. Benzene has also been found to be teratogenic in rats [86].

Methylene dichloride (dichloromethane) when tested by inhalation in mice, rats and hamsters, increased the incidences of benign and malignant lung and liver tumours in mice of each sex, the incidence and multiplicity of benign mammary tumours in rats of each sex and the incidence of sarcomas located in the neck of male rats [87].

Oral administration of methylene dichloride resulted in an increased incidence of lung tumours in male mice and an increased incidence of malignant mammary tumours in female rats [45].

The evidence of carcinogenicity of methylene chloride to animals has been judged by IARC to be *sufficient*, while the evidence for carcinogenicity to humans is *inadequate* [45].

Methylene chloride is mutagenic to bacteria, induces sex-linked recessive lethal mutations in *Drosophila* [64]. It induced chromosomal aberrations, but not mutation or DNA damage in rodent cells *in vitro* [45, 64].

Trichloroethylene produced hepatocellular carcinomas and lung tumours in both male and female mice following oral administration. One oral study in rats was considered to be inadequate, and the other showed equivocal evidence of carcinogenicity [45, 88]. Inhalation studies with trichloroethylene have been carried out in mice, rats and hamsters with mixed results. In one study with female mice it caused lung tumours, but gave negative results in the other study in mice and in rats and hamsters [45].

In mice, oral administration of trichloroethylene containing the stabilizer epichlorohydrin induced forestomach carcinoma, but no liver or lung carcinoma [89]. When pure trichloroethylene was tested by oral administration in mice and rats, hepatocellular carcinomas were found in male and female mice but not in female rats. The study in male rats was considered inadequate [90]. IARC considers the evidence for carcinogenicity of trichloroethylene to animal to be *limited*, and the evidence for carcinogenicity to humans to be *inadequate* [45].

Trichloroethylene induced micronuclei, somatic mutation (in the spot test), sperm anomalies and DNA strand breaks in the kidney and liver, but not lung, of mice treated *in vivo*. It did not induce dominant lethal mutations. It induced SCE's and unscheduled DNA synthesis in human lymphocytes *in vitro*. Additionally trichloroethylene induced transformation of mouse and rat cells but not of Syrian hamster cells and did not induce SCE's in Chinese hamster cells *in vitro* or unscheduled DNA synthesis in rat hepatocytes [45, 64]. It should be noted that many *commercial* preparations of trichloroethylene contain stabilizers such as epichlorohydrin and in many genotoxic studies the *purity* of the preparations tested are not given [45].

The evidence for the carcinogenicity of 1,1,1-trichloroethane to experimental animals is considered *limited* [3, 45] 1,1,1-trichloroethane is mutagenic in *E. Coli* in the presence of liver microsomal activation [3, 91]. Dimethylformamide (DMF) is metabolized in rats to N-methylformamide [92] which is teratogenic and embryotoxic to rats [93].

Methyl ethyl ketone (MEK) was embryotoxic, fetotoxic and potentially teratogenic following administration by inhalation to pregnant rats [94].

Ethylene dichloride is carcinogenic to mice and rats and is mutagenic to *S. typhymurium* with or without metabolic activation [88].

e) *Fillers and Extenders*

Carbon black is the principal filler of toxicological concern although it is recognized that an unambiguous assessment is difficult since different preparations contain variable amounts of many compounds [3, 95–97].

Carbon black is second only to elastomers in distribution and importance within the rubber industry. About 90% of the more than 10 billion pounds of carbon black produced each year is used in rubber products [96, 97]. The most probable carcinogenic hazard is acknowledged to be associated with the "benzene extractables", principally from furnace black. These extractables are composed mainly of aromatic hydrocarbons and sulphur compounds. Among the polycyclic aromatic hydrocarbons (PAH's) that have been identified in benzene extractables, are the carcinogenic PAH's benzo(a)pyrene, benz(a)anthracene and indeno (1,2,3-*cd*) pyrene [3, 95]. Additionally chrysene has been described as an initiating agent and pyrene and fluorenthene as cocarcinogenic for mouse skin [3, 95]. Existing data would suggest that the known carcinogens present in carbon blacks are strongly absorbed but can be eluted by biological fluids [3].

Mineral oils (e.g., coal-tar oils, petrolatum) and tar products such as bitumen and pitch are widely used in the rubber industry as extenders [3, 9, 98]. Since the mineral oils are comparatively inexpensive and provide desirable properties to the finished rubber, their amount in tires has increased over the past year to about 20% or even more [3]. It is broadly acknowledged that extenders such as the high-boiling mineral oil distillates (also referred to as aromatic oils) obtained from residues of solvent-refining and the manufacture of lubricating and curting oils contain relatively large amounts of PAH's and 30% and more of the oil can consist of 4–6 ring PAH's [3, 97]. Additionally, PAH's may also be formed when tars and mineral oils are heated [3, 97].

Analogous to carbon blacks, mineral oils and tar products vary in composition depending on their origin, methods of production, etc. Mineral oils and tar products all induce carcinogenic effects in mammals including man [3, 96]. The carcinogenicity of these components is believed to be dependent on the presence of carcinogenic PAH's, most prominent of which is benzo(a)pyrene. Additionally, benzo(b)fluor-anthene, benzo(j)fluoranthene, chrysene, dibenz(a,h)anthracene, dibenzo(a,b)pyrene, dibenzo(a,i)pyrene and indeno(1,2,3-cd)pyrene have also been considered by an IARC Working Group [3, 99].

Mineral oils containing different amounts of PAH's have also been found to be mutagenic in the Ames *Salmonella*/microsome assay [3, 57, 100].

f) *Nitroso Compounds*

The formation of nitroso derivatives from rubber chemical precursors, as well as their presence as contaminants in some rubber chemicals, the potential for nitrosation *per se* of a number of rubber chemicals have all been noted earlier in this chapter. It should be restated that N-nitroso derivatives (e.g., di-n-butyl-, diethyl-; dimethyl-; morpholine-; piperidine) have all been shown to be potent carcinogens in a variety of animal species often administered by various routes [3, 21, 45]. These compounds are additionally metabolized to reactive metabolites which are mutagenic in a variety of genotoxicity test systems [3, 21, 45, 64].

Exposure in the Rubber Industry

A number of general reviews of exposure in the rubber industry should be noted [3]. The principal reviews include: 1) comprehensive reviews of levels of exposure to particulates and solvents in the U.S. tire manufacturing industry [101, 102], 2) environmental monitoring study in 10 U.K. tire factories in 1974 [103], 3) sampling and identification of volatile pollutants in Italian rubber manufacturing processes [104], 4) studies of nitrosamine exposure in the U.S. facilities [16, 105], and facilities in the Federal Republic of Germany [21].

The identification of volatile pollutants in an Italian shoe factory, tire retreading factory and electrical cable insulating plant is particularly revealing [104]. Table 5 lists the chemicals employed in the three rubber manufacturing plants, while table 6 lists the compounds identified by gas-chromatography/mass spectrometry (GC/MS) in these plants and observed environmental concentration ranges. Approximately 100 different compounds were identified and quantitated with their cumulative concentrations ranging from 25 to 27000 micrograms/cubic meter. Table 7 lists the chemical families of identified compounds, the number of observed compounds per family and the measured range of environmental concentrations ($\mu g/m^3$). It is of interest to note the wide distribution of the pollutants between chemical families (Table 7) some of which were not present in the raw materials, e.g., ethers, aldehydes, alcohols, ketones and some sulfur — and nitrogen containing compounds. The above study of Cocheo et al. [104] indicated that a number of pollutants identified in the plants are of toxicological concern. It is reasonable to suggest that at least some of them may contribute in some manner to the elevated morbidity found among the workers employed in the rubber products industry [4].

Table 5. Chemicals Employed in Rubber Manufacturing Plants

Ref. No.	Common Name (CAS Reg. No.)	Users		
		A[a]	B[b]	C[c]
	Polymers			
1	Natural rubber (polyisoprene) (9003-31-0)	x	x	
2	SBR (styrene-butadiene copolymer) (9003-55-8)	x	x	x
3	cis-Polybutadiene		x	
4	EVA (ethylene-vinyl acetate copolymer) (24937-78-8)			x
5	Acrylonitrile-vinyl chloride copolymer (9003-00-3)			x
6	Polychloroprene (9010-98-4)			x
	Vulcanizing Agents			
7	Sulfur (7704-34-9)	x	x	
8	Tetramethylthiuram sulfide (97-91-6)	x	x	x
9	Dicumyl peroxyde (80-43-3)			x
	Accelerators			
10	Diphenylguanidine (103-06-7)	x	x	
11	2-Mercaptobenzimidazole (583-39-1)	x		
12	2-Mercaptobenzothiazole (149-30-4)	x	x	
13	N-Cyclohexyl-2-benzothiazolylsulfenamide (95-33-0)	x	x	x
14	2-Mercaptobenzothiazole disulfide (120-78-5)	x		
15	Piperidine pentamethylenedithiocarbamate (98-77-1)	x		
16	Zinc N-ethyl-N-phenyldithiocarbamate (14634-93-6)	x		
17	2-(N-Morpholinyl)-mercaptobenzothiazole (102-77 2)		x	
18	Hexamethylenetetramine (100-97-0)	x	x	
19	Ethylene thiourea (96-45-7)			x
	Activators			
20	Zinc oxide (1314-13-2)	x	x	
21	Zinc carbonate (10476-83-2)	x	x	
22	Magnesium oxide (1309-48-4)			x
23	Stearic acid (57-11-4)	x	x	x

Table 5. (continued)

Ref. No.	Common Name (CAS Reg. No.)	Users		
		A[a]	B[b]	C[c]
	Antiozonants			
24	2.2.4-trimethyl-1,2-dihydroquinoline (polymer) (147-47-7)	×	×	
25	N,N'-(1,3-dimethylbutyl)-p-phenylenediamine (793-24-8)		×	
26	Phenyloctylamine (mixture of isomers)			×
27	N,N'-ditolyl-p-phenylenediamine (mixture of isomers)			×
28	Paraffinic wax		×	×
	Antioxidants			
29	Alkylphenols	×	×	
30	Resorcinol (108-46-3)		×	
31	Bisphenol A (80-05-7)	×		
32	2,6-Diterbutylhydroquinone (2444-28-2)	×	×	
	Retarders			
33	Benzoic acid (65-85-0)	×		
34	Phthalic anhydride (85-44-9)	×		
35	N-Cyclohexylthiophthalimide (17796-82-6)	×	×	
	Plasticizers			
36	Aliphatic oil	×	×	×
37	Aromatic oil	×	×	×
38	Naphthenic oil	×	×	×
39	Di-(2-ethylhexyl)-phthalate (117-81-7)	×	×	×
40	Coumarone-indene resin	×		
41	Rape seed oil	×		
42	Ethylene dimethacrylate (97-90-5)			×
	Blowing Agent			
43	Benzenesulfonylhydrazide (80-17-1)	×		
	Extenders			
44	Calcium carbonate (471-34-1)	×		×
45	Silica gel (7631-86-9)	×	×	×
46	Barium sulfate (7727-43-7)	×		
47	China clay	×		×
48	Carbon black (7440-44-0)	×	×	×
	Lubricant			
49	Zinc stearate (557-05-1)			×

[a] Shoe-sole factory; [b] Tire retreading factory; [c] Electrical cable insulating plant

Epidemiology in the Rubber Industry

There are many published reports of large scale epidemiology studies in the rubber industry with many of these being retrospective mortality studies [3, 106–123]. Although the association between occupational exposures to chemicals and malignant tumours of the lower urinary tract was first suspected in the German dyestuffs industry in 1895, until the late 1940's no suspicion existed that bladder tumours

Table 6. Compounds Identified in Rubber Manufacturing Plants and Observed Environmental Concentration Ranges

Compound	CAS Reg. No.	R.R.T.[a]	Observed Concentration Range (μg/m³)				Probable Source (See Table 5)
Sampling Site Number of Samples			A[b] 13	B1[c] 6	B2[d] 6	C[e] 10	
Benzene	71-43-2	0.29	8-15	10-1200	25-180	N.D.[f]	37
Cyclohexene	110-83-8	0.30	3-12	N.D.	N.D.	N.D.	1-3
Methylcyclohexane	108-87-2	0.34	1-2	3-800	3-25	N.D.	38
Toluene	108-88-3	0.43	4-8	6-800	20-160	0-150	37
Ethylcyclopentane	1640-89-7	0.44	50-180	N.D.	N.D.	N.D.	38
Ethylcyclohexane	1678-91-7	0.53	4-9	N.D.	N.D.	N.D.	38
Heptane	142-82-5	0.54	20-1400c	3-500	0-70	N.D.	36
4-Vinylcyclohexene	100-40-3	0.66	30-210	N.D.	0-3	0-10	1-2-3
Ethylbenzene	100-41-4	0.6	30-150	2-90	1-15	0-30	3
Octane	111-65-9	0.77	0-300	6-90	0-10	0-1	36
p-Xylene	106-42-3	0.79	15-40	7-120	2-10	0-110	37
Cyclohexanone	108-94-1	0.83	0-10	N.D.	N.D.	N.D.	40
Styrene	100-42-5	0.87	90-500	2-180	1-20	0-5	2
Nonane	111-84-2	0.93	0-60	0-10	0-1	0-5	36
1-Ethyl-4-methylcyclohexane[g]		0.97	1-2	N.D.	N.D.	N.D.	38
α-Pinene[h]	80-56-8	1.03	N.D.	N.D.	N.D.	0-5	38
1,4-Cyclohexadiene-1-isopropyl-4-methyl[h]	99-85-4	1.12	0-50	2-10	0-5	N.D.	1
Isopropylbenzene	98-82-8	1.13	60-250	2-200	0-10	N.D.	37
Cyclohexene-1-methyl-3-(1-methylvinyl)[h]	499-03-6	1.19	N.D.	0-50	N.D.	N.D.	1
Isopropylcyclohexane	696-29-7	1.22	1-3	N.D.	N.D.	N.D.	38
Propylbenzene	103-65-1	1.24	30-300	0-15	0-2	N.D.	37
Benzaldehyde	100-52-7	1.28	N.D.	0-6	N.D.	0-3	2
α-Methylstyrene	98-83-9	1.31	N.D.	N.D.	N.D.	0-5	2
1-Isopropyl-4-methylcyclohexane (trans)	1678-82-6	1.36	4-400	1-100	0-15	N.D.	38
1-Isopropyl-4-methylcyclohexane (cis)	6069-98-3	1.38	4-300	1-80	0-10	N.D.	38
1,4-Diethylbenzene	105-05-5	1.40	N.D.	N.D.	N.D.	0-2	37
1-Isopropyl-3-methylcyclohexane[g]		1.45	N.D.	1-70	0-10	N.D.	38
Ethylene diacetate	111-55-7	1.53	N.D.	N.D.	N.D.	0-5	42
Decane	124-18-5	1.55	1-370	0-20	0-2	0-20	36

Triisobutylene[h]	7756-94-7	1.58	300-7000	5-180	0-20	0-3	2-3-38
Cyclohexene-5-methyl-3-(1-methylvinyl)[h]	65816-08-1	1.59	N.D.	2-370	0-20	N.D.	1
Indane	496-11-7	1.64	15-35	0-5	N.D.	N.D.	38
1-Isopropyl-4-methylbenzene	99-87-6	1.68	N.D.	5-450	1-20	N.D.	37
2,2,4,6-Pentamethylheptane	13475-82-6	1.70	N.D.	N.D.	N.D.	0-10	36
Cyclohexene-1-methyl-4-(1-methylvinyl)	138-86-3	1.72	25-130	5-1700	0-10	N.D.	1
4-ter-Butylcyclohexene[h]	2228-98-0	1.78	3-5	N.D.	N.D.	N.D.	1
1-Isopropyl-2-methylbenzene	527-84-4	1.91	N.D.	0-2	1-15	N.D.	37
Acetophenone	98-86-2	1.93	N.D.	N.D.	N.D.	0-30	9
Dimethylstyrene[i]		2.07	N.D.	1-70	0-20	N.D.	2
2-Phenylisopropanol[h]	617-94-7	2.10	N.D.	N.D.	N.D.	0-25	9
Undecane	1120-21-4	2.22	1-170	0-35	0-3	0-3	36
Tetramethylbenzene[i]		2.28	0-120	0-15	N.D.	N.D.	37
1,2,3,4-Tetrahydronaphthalene	119-64-2	2.31	N.D.	0-1	N.D.	N.D.	38
2-(3-Cyclohexenyl)-acetaldehyde[j]	24480-99-7	2.48	0-30	N.D.	N.D.	N.D.	2
1,3-Diisopropylbenzene	99-62-7	2.56	N.D.	1-40	0-5	0-2	37
1,2-Diisopropylbenzene	577-55-9	2.66	10-20	N.D.	N.D.	N.D.	37
1,4-Diisopropylbenzene	100-18-15	2.69	25-50	N.D.	0-2	0-1	37
2-Isopropyl-6-methylphenol[h]	89-83-8	2.80	N.D.	0-35	1-70	N.D.	29
2-Butanol-4-isopropoxy[j]	40091-57-4	2.85	N.D.	2-320	N.D.	0-3	9
Cyclohexene-1-chloro-5-(1-chlorovinyl)[h]	13547-07-4	2.96	N.D.	N.D.	0-5	0-5	6
Cyclohexene-1-chloro-4-(1-chlorovinyl)[h]	13547-06-3	3.01	N.D.	N.D.	0-2	0-2	6
Cyclohexylisothiocyanate[j]	1122-82-3	3.08	0-100	1-335	N.D.	N.D.	35
2,2,4-Trimethyl-1,2-dihydroquinoline[j]	147-47-7	3.15	0-15	N.D.	N.D.	N.D.	24
Cyclododecatriene[i]		3.26	N.D.	5-400	0-10	N.D.	3
Dodecane	112-40-3	3.31	1-60	0-25	0-3	0-10	36
Chloromethylbenzoate[j]	5335-05-7	3.38	N.D.	N.D.	N.D.	0-20	?
Tridecane	629-50-5	3.54	N.D.	1-130	1-5	0-4	36
Tetraisobutylene[h]	15220-85-6	3.60	10-300	1-60	0-8	N.D.	2-3-38
p-ter-Butylstyrene	1746-23-2	3.60	N.D.	0-110	0-2	N.D.	2
Ethylene dimethacrylate	97-90-5	3.66	N.D.	N.D.	N.D.	0-50	42
1,5-Cyclooctadiene-1,6-dichloro[h]	29480-42-0	3.75	N.D.	N.D.	N.D.	0-10	6
Dimethylpropylhexahydronaphthalene[i]		4.00	0-220	0-83	0-1	N.D.	38
Tetradecane	629-59-4	4.15	0-40	0-70	0-3	0-5	36
Octylbenzene[i]		4.29	8-15	N.D.	N.D.	N.D.	37
Nonylbenzene[i]		4.41	0-240	0-25	N.D.	N.D.	37

Table 6 (continued)

Compound	CAS Reg. No.	R.T.[a]	Observed Concentration Range (µg/m³)				Probable Source (See Table 5)
Sampling Site Number of Samples			A[b] 13	B1[c] 6	B2[d] 6	C[e] 10	
2,6-Di-ter-butyl-p-quinone	719-22-2	4.52	30-170	0-140	0-1	N.D.	32
Heptylphenol[i]		4.69	0-150	N.D.	N.D.	N.D.	29
Pentadecane	629-62-9	4.74	0-30	1-80	0-3	0-12	36
1,6-Dimethyl-4-isopropyl-1,2,3,4-tetra-hydronaphthalene[h]	483-77-2	4.83	0-60	0-30	N.D.	N.D.	38
Octylphenol[h]		4.93	0-40	N.D.	N.D.	N.D.	29
Decylbenzene[i]		4.98	N.D.	0-30	N.D.	0-1	37
Di-ter-butylthiophene[i]		5.02	0-500	0-80	N.D.	N.D.	37
1,1-Dimethyl-5-ter-butylindane[h]	38393-97-4	5.19	N.D.	N.D.	N.D.	0-1	37
Ether isopropyl-2-phenylisopropyl[j]	not found	5.25	N.D.	N.D.	N.D.	0-1	9
Diethylphthalate	84-66-2	5.26	0-120	0-30	0-1	1-3	39
Hexadecane	544-76-3	5.29	1-170	0-60	0-2	0-10	36
Benzophenone	119-61-9	5.39	N.D.	N.D.	N.D.	0-1	9
Dodecylbenzene[i]		5.60	10-400	N.D.	N.D.	N.D.	37
1,2-Di-tolyethane[h]	538-39-6	5.68	N.D.	N.D.	0-1	N.D.	2
Heptadecane	629-78-7	5.82	0-200	0-110	0-2	0-10	36
1-Phenyl-1,3,3-trimethylindane[h]	3910-35-8	5.8	8-15	N.D.	N.D.	N.D.	37
2,6-Di-ter-butyl-4-ethylphenyl	4130-42-1	6.13	0-200	2-420	0-3	N.D.	29
Dodecylphenol[i]		6.24	0-4	N.D.	N.D.	N.D.	29
Octadecane	593-45-3	6.31	0-200	1-65	0-2	1-10	36
Cumyl hydroperoxide	80-15-9	6.39	N.D.	N.D.	N.D.	0-60	9
1-Phenylnaphthalene[h]	605-02-7	6.51	N.D.	0-5	0-1	0-2	37
Nonadecane	629-92-5	6.61	N.D.	N.D.	N.D.	0-6	36
Diisobutyl phthalate	84-69-5	6.67	N.D.	10-2500	4-15	N.D.	39
Tridecylbenzene[i]		6.81	70-360	1-150	N.D.	N.D.	37
Diphenylbenzene[h]	26140-60-3	6.88	N.D.	N.D.	N.D.	0-1	37
Dibutyl phthalate	84-74-2	7.05	0-150	5-500	N.D.	0-35	39
Eicosane	112-95-8	7.22	N.D.	0-10	0-1	0-3	36
Tetradecylbenzene[i]		7.37	50-280	N.D.	N.D.	N.D.	37

N-Cyclohexylmercaptobenzimidazole[j]	68406-57-5	7.60	0-700	N.D.	N.D.	N.D.	11-13
Heneicosane	629-94-7	7.60	0-700	N.D.	N.D.	N.D.	11-13
Heptadecylphenol[i]		7.79	0-5	N.D.	N.D.	N.D.	29
Docosane	629-97-0	8.05	N.D.	0-6	N.D.	N.D.	36
Di-(2-ethylhexyl) phthalate	117-84-0	9.34	N.D.	0-2	N.D.	N.D.	39
Farnesol[h]	4608-84-0	10.29	0-25	N.D.	N.D.	N.D.	1

a Retention time relative to internal standard (ethylene glycol ethyl ether acetate).
b Shoe-sole factory, vulcanization area.
c Tire retreading factory, vulcanization area.
d Tire retreading factory, extrusion area.
e Electrical cables insulation plant, extrusion area.
f Not detected.
g Tentatively identified compounds (standard not available), good match with library spectrum, but unidentified cis-trans isomer.
h Tentatively identified compounds (standard not available), good match with library spectrum.
i Tentatively identified compounds (standard not available), good match with library spectrum, but unidentified isomer.
j Tentatively identified compounds (standard not available), spectrum not found in library.

Table 7. Chemical Families of Identified Compounds: Number of Observed Compounds, Mean and Range of Environmental Concentrations (µg/m³)

Sampling Site Number of Samples	A[a] 13			B1[b] 6			B2[c] 6			C[d] 10		
Family (No. of Comp.)	No. of Comp.	Concentration		No. of Comp.	Concentration		No. of Comp.	Concentration		No. of Comp.	Concentration	
		Mean	Range		Mean	Range		Mean	Range		Mean	Range
Alkanes (16)	9	3560	20-14200	13	190	7-920	11	33	2-80	13	26	4-70
Cycloalkanes (11)	10	2120	350-8100	7	304	1-1220	7	24	4-58	2	1	5-5
Cycloalkenes (10)	6	138	0-830	6	485	0-2600	6	17	0-30	2	5	0-25
Aromatic hydrocarbons (32)	20	605	0-3130	23	813	16-3730	16	187	58-420	13	50	0-300
Chlorinated (4)	0	—	—	0	—	—	0	—	—	4	4	0-11
Phenols (6)	6	73	0-230	2	178	4-860	2	21	0-86	0	—	—
Esters (6)[e]	3	70	0-200	4	545	10-3000	2	10	5-15	4	32	1-85
Miscellaneous (14)[e]	6	165	0-800	3	95	1-475	1	0.2	0-1	7	37	0-145
TOTAL (99)	60	6700	960-27000	58	2700	60-13000	45	300	100-490	45	140	25-550

[a] Shoe-sole factory, vulcanization area.
[b] Tire retreading factory, vulcanization area.
[c] Tire retreading factory, extrusion area.
[d] Electrical cables insulation plant, extrusion area.
[e] Ethers, sulfur- and nitrogen-compounds, aldehydes, alcohol, ketones, peroxides and quinones.

could result from exposure to rubber chemicals [3]. It was not until 1954 that Case and Hosker [124] identified an excess of bladder tumours affecting British rubber workers exposed to compounding chemicals which were withdrawn from use in the U.K. in 1949. Cancer of the urinary bladder, clearly excessive in British workers employed before 1950, particularly those in jobs likely to entail exposure to aromatic amines, appears not to be increased in employes who entered the rubber industry in the U.K. since that date [3]. The withdrawal in 1949 of certain antioxidants containing 1-naphthylamine from the U.K. rubber industry appears likely to have accounted principally for the subsequent decline in bladder cancer [3].

Among rubber workers in the U.S., excess malignancies of the lymphatic and haematopoietic systems (particularly lymphatic leukemia) have been associated with jobs entailing exposure to solvents [3, 106, 107, 109, 110, 113, 114]. Benzene, recognized as a human carcinogen [3, 45] was once extensively used as a solvent within the rubber industry but may still be present as a contaminant of other organic solvents such as toluene [3].

Elevated incidences of stomach cancer reported in U.S. and British rubber workers appears to be associated with jobs such as compounding and mixing, milling and extrusion which are early in the production line of rubber products manufacture [3].

Although lung cancer is positively related to a variety of jobs within the rubber industry, it is not possible to attribute specific factors in the job category to date [3].

Table 8 summarizes the strength of currently available evidence for past or present cancer excesses in the rubber industry [3]. IARC considers that the evidence for five types of cancer (bladder, stomach, lung, skin and leukemia) is either

Table 8. Strength of currently available evidence for past or present cancer excesses in the rubber industry

Strength of evidence	Type of cancer	Presumed agent or job category
Sufficient for excess occurrence in rubber workers and for causal association with occupational exposures	Bladder Leukaemia	Aromatic amines Solvents
Sufficient for excess occurrence in rubber workers; and *limited* for causal association with occupational exposures	Stomach Lung	Compounding, mixing and milling Various
Limited for excess occurrence in rubber workers and for causal association with occupational exposures	Skin	Tyre building
Limited for excess occurrence in rubber workers; and inadequate for causal association with occupational exposures	Colon Prostate Lymphoma	
Inadequate for excess occurrence in rubber workers and for causal association with occupational exposures	Brain Thyroid Pancreas Oesophagus	

"limited" or "sufficient" for excess occurrence in rubber workers and for causal association with occupational exposure [3].

In addition to potential for carcinogenic risk, numerous studies with exposure indicators, such as mutagenic activity in urine, thioether extraction and sister chromatid exchange strongly suggest that workers in the rubber industry are exposed to mutagens [3, 115].

It is important to note the conclusions of IARC [3], that the combination of chemical exposures which occur in the rubber industry is probably more relevant to the observed cancer pattern than are single compounds or groups of compounds. The variety of exposures in this industry to a galaxy of chemicals, increases the probability that there are interactive effects between two or more such agents and hence potentially increasing the toxicological risk associated with such exposures.

Additional Toxicological Areas of Concern

While this chapter has principally focused on the carcinogenic, genotoxic and terato-genic effects of the major monomers and rubber additives, it is important to stress other potentially significant toxicides to these agents. These include: eye, skin and pulmonary irritation, skin sensitization, central nervous system effects, neurotoxicity, issue changes and immunosuppressive effects. Of these toxicological effects, by far the most common effects, are those of eye and skin irritation and skin sensitization, which have been reported to result from exposure of contact to literally every class of rubber chemical including: curing agents, accelerators, activators, retarders, blowing agents, antitack agents, plasticizers, solvents, antioxidants and antiozono-nants [2].

For example, eye, skin irritation and skin sensitization have resulted from exposure to: 1) *curing agents* (e.g., selenium and sulfur derivatives, thiuram di-sulfides, phenylenediamines); 2) *accelerates* (e.g., metallodinlykedithio carbonates, aryl guanidines and biguanides, thiocarbamides, and mercaptobenzothiazoles; 3) *activators* (e.g., cyanurates and alkoxysilanes; 4) *antioxidants* and *antiozonants* (e.g., phenylenediamines, quinolines, cresols, and arylamines); 4) *retarders* (e.g., anhydrides, thiophthalimides); 5) *antitack agents* (e.g., metal stearates, talc); 6) *plasticizers* (e.g., phthalates, adipates, azelates); 7) *bonding agents* (e.g., di- and triisocyantes; 8) *blowing agents* (e.g., 1,1-azobisformamide).

A number of solvents are neurotoxic per se (e.g., hexane, toluene) and others such as methyl ethyl ketone (MEK) are known to potentiate [125] the toxicities of other organic solvents such as the neurotoxicity of n-hexane [126, 128].

Monomers such as acrylonitrile, styrene and butadiene have been associated with a variety of central nervous system, neurotoxic and immunosuppressive effects [34, 41, 129, 130].

Neurotoxic effects have been established for a number of dithiocarbamates, thiram and thiuram sulfide accelerators in both mammaliam and avian species [131].

Disposal and Reclaiming of Rubber

Enormous quantities of tires are globally disposed of annually. The United States alone generates approximately 500×10^6 scrap tires per year; approximately 67%

of the scrap rubber, primarily as tires was used as landfill in 1977 [132, 133], although this method of tire disposal is decreasing in the U.S. due to the general decrease in landfill usage *per se* for disposal. There are a number of recognized difficulties involved in the disposal of tires in landfills which include: 1) migration of tires that are whole to the surface in a landfill due to entrapment landfill-generated gas; 2) possibility of fire resulting in the emission of a variety of toxic substances (e.g., PAH's., aromatic hydrocarbon and chlorinated solvents) metals, carbon particles or soot) and 3) possible eventual leaching of rubber chemicals and their degradation products into aqueous waste streams or ground water.

Although in the past, it has been generally uneconomical to pyrolyze rubber scrap because of uncompetitive costs and few product markets, high fuel costs and petroleum scarcity in Europe and other parts of the Eastern Hemisphere make rubber reuse as a fuel source more economical there than in the United States [133]. The use of scrap rubber for fuels appear to be one of the best alternatives for reusing rubber as natural gas and fuel-oil costs increase. Tires that contain more than 90% organic materials have a heat value of about 14,000 Btu/lb., compared to coal (about 8000–12,000 Btu/lb.) [133].

Scrap-tire pyrolysis has been the subject of a number of research studies by rubber, oil and carbon-black interests throughout the world [133]. In one procedure known as the Tosco II process, chopped tires are fed into a rotary drum with hot ceramic balls at 480–549 °C in a reducing atmosphere. The rubber pyrolyzes and forms a solid residue, an oil vapor and off-gases. The off-gas is basically a combination of ethylene, propylene and butylene. The oil, which contains about 1% sulfur can be substituted directly for fuel oil. The production of carbon blacks and other related contents in the residue from tire pyrolysis (e.g., ZnO, possibly other inorganic materials and glass fiber, from old tires are however unpredictable and hence the residue mixture is basically useful only as low grade filler for mechanical goods and is uncompetitive as a carbon black source [133].

In another study, the U.S. Bureau of Mines-American Gas Association pilot plant, elaborated the products of rubber tires pyrolysis conducted at temperatures ranging from 500 °C to 900 °C [134, 135]. With increasing pyrolysis temperature, residues increased, the yield of heavy oil decreased and the amount of gas increased. The (volume%) of the components found in the off-gas from 500 °C pyrolysis were hydrogen (51.8), methane (18.9), ethane (18.7), propane (5.3), propylene (3.0), isobutane (0.5), butane (0.5), butene-1 (0.2), isobutylene (0.9), *trans*-butene-2 (0.1) and *cis*-butene-2 (0.1). The heavy oil contained alkylbenzenes and alkylnaphthalenes as the predominant compounds determined by spectrophotometric analysis. The *probable* compounds identified from mass spectra included alkylbenzenes, phenols, 3-ring aromatics, biphenyl, acenaphthene, alkyl-naphthalenes, 3-ring aromatics, indenes, styrene, alkylstyrenes and/or indans. Pyrolysis studies demonstrate that pyrolysis of scrap rubber yields a complex mixture of liquids, gas and residue in varying ratios dependent on the nature of the scrap rubber and the conditions of pyrolysis [133–135].

A major use for scrap tires is in the production of asphalt-rubber for roadways. Production of asphalt rubber has increased from 450 tons in 1970 to 30,000 tons in 1980. In 1980, the U.S. used about 4,500 tons of rubber in rubber asphalt which is less than 5% of the recycled rubber produced. If it is assumed that rubber asphalt contains about 25% rubber and 75% asphalt, potential demand for scrap rubber is about 45,000 tons/year or about 2% of the amount available [133].

One conventional method for mixing asphalt rubber is to mix the reclaimed and crumb rubber and asphalt to about 175–220 °C for 1–2 hours. (The rubber crumb is usually scrap tires that are ground into particles that are less than 2 mm in diameter). Besides the use of asphalt rubber for roadways, it is used for roofing materials, crack-and-joint sealers, hot-mix binders and winterproofing membranes. The rubber improves asphalt ductility increases the temperature at which asphalt softens and increases the long-term durability of asphalt [132].

By far, the most important source of raw material for rubber reclaiming is tire scrap. The quality of rubber in tires is high, giving an unusually high percentage of rubber hydrocarbon at low cost. There are 3 major steps to the reclaiming process: the vulcanized rubber scrap is first ground, then given a heat treatment for depolymerization and finally processed by intensive friction milling or pelletizing [133, 136]. Reclaimed rubber can be used by itself in adhesives, and solid-rubber-tire compounds, or it can be blended with virgin rubber. Compounding is based on rubber hydrocarbon content (RHC). Until the 1960's reclaimed rubber was a principal raw material in many molded and extruded rubber goods, e.g., tires, rubber mats and hard-rubber battery cases. The advent of vinyls and other plastics and less expensive oil-extended synthetic polymers and increased radial-tire production has diminished the utility of reclaimed-rubber compounds.

The main sources of scrap and reclaim rubber are tires, natural-rubber tubes and butyl tubes. Speciality reclaims are made from scrap silicone, chloroprene (CR), nitrile-butadiene (NBR) and ethylene-propylene-butadiene-*ter*-polymer (EPDM) rubber scraps. Tires, hoses, belts, molded and extruded goods and asphalt products consume about 80% of the reclaim rubber manufactured [133]. Roughly 75% of the discarded tires are disposed of in landfills; 20% are retreaded and 5% are reclaimed, burned for fuel, split etc. Scrap whole tires have also been used for artificial fishing reefs, oyster beds and as a floating break water [133].

Emission and Degradation of Rubber Chemicals

Initially, it should be noted that the monomers used in the rubber industry (Table 4), due to their volatility, would be the class of agent most likely to be released to the environment via emissions from vents on process equipment, storage tank losses, miscellaneous leaks and spills, process waste waters and solid process wastes.

The emissions resulting from the curing and/or vulcanization processing in the compounding or formulation of rubber are dependent on the temperature, physical and structural properties of the various compounding ingredients [3, 137]. General purpose polymers maintain their integrities in the presence of heat until pyrolylis temperaturs (200 °C–400 °C) are reached. Only upon continued heating at 175–250 °C for several hours have depolymerization process been noted [138]. Curing operations of shorter duration should result in little or no breakdown. Thus polymeric losses will be a result of residual amounts of monomer and impurities from manufacturing processes and should represent less than 1% of polymer on a weight basis

[137]. Monomers however, are sufficiently volatile to undergo appreciable loss in prevulcanization milling and calendering operations, (e.g., acrylonitrile, b.p. 78 °C, chloroprene, b.p. 59 °C and styrene, b.p. 145 °C).

In regard to antioxidants and antiozonants, in most cases heating losses of phenolic compounds are higher than for their amine counterparts [137, 139, 140]. Normal vaporization in curing rubber generally results in a weight reduction of 0.5–0.1% of antidegradant present.

Common accelerators such as the dithiocarbamates, thiurams, sulfenamides, thiazoles and guanidines have melting points in the range of 70°–200 °C and thus some loss at curing temperatures would be expected [137]. For example, accelerators such as mecaptobenzothiazole (MBT) derivatives (e.g., sulfenamides) have a strong tendency to decompose in the vulcanization procedure and can form MBT which may carry over in to end products and waste streams [141].

Vulcanizing agents and retarders are most likely to be volatilized at process temperatures. These substances include a variety of amines, esters and organoacids which are either liquids at room temperature or solids with melting points between 70° and 200 °C. Losses of 1% might be expected [137].

It is important to stress that the materials incorporated in the formulation are invariably formulations of technical grade. The purity of the main isomer or ingredient is thus generally low about 60–95%. Some impurities will possess sufficient volatility to be released. Gas chromatographic analysis of commercial antioxidants has confirmed high levels of impurities [142]. Thus overall it would be a reasonable assumption to conclude that a large number, perhaps hundreds of impurities can be emitted into workroom air and potentially into the general environment via air and waste water emissions [137].

The degradation of synthetic oligomers by microorganisms has been reported by Tsúchii et al. [143]. For example, *acinetobactor* species 351 from soil grows well on butadiene oligomers and degrades 30% of commercial samples in 3 days. In a study of the biodegradability of rubber additives by soil microorganisms, William [1] reported that the majority of derivatives e.g., accelerators and vulcanizing agents such as arylguanidines did not support microbial growth and were highly toxic to soil microorganisms even at very low concentrations.

Aniline and nitrobenzene are used extensively in the production of a variety of vulcanization accelerators and antioxidants. They are discharged in the aqueous waste which may subsequently accumulate in the environment and prove toxic to many organisms [145]. Aniline and nitrobenzene in aniline plant waste water were found to be biodegradable in the presence of each other as determined by gas chromatography [146].

A large number of both inorganic and organic rubber chemicals and their degradation products have been found in many hazardous waste sites in the U.S. and elsewhere. Since the majority of these sites in the U.S. are considered uncontrolled hazardous waste sites which include material from diverse sources such as byproducts of rubber manufacture, liquid waste processing and incineration and pesticide production, it is extremely difficult to unambiguously ascribe a specific chemical entity as solely arising from one waste disposal process for example as in the case of a variety of chlorinated and aminoaryl derivatives [148].

Analytical Methodology

While a variety of instrumental analytical techniques have been employed for the determination of individual components and in production methodologies in the rubber industry, the principal methodologies rely on gas chromatographic (GC) and spectrophotometric techniques. Pyrolysis gas chromatography, optimally coupled with mass spectrometry has proven to be especially effective in the analysis of the monomeric constituents in synthetic elastomers and their blends with natural rubber or in various mixtures of elastomers. For example, pyrolysis GC has been employed for the identification of polymers such as polyethylene, isoprene and butadiene rubbers; butadiene copolymers; nitrile-, butyl-, polyisobutylene, EPDM-, and silicone rubbers as well as butylnatural rubber blends using individual products to characterize the compounds and structures of the macromolecules [18–157].

A high variety traditional gas chromatographic procedures have been employed for the analysis of various rubber chemicals, e.g., ester plasticizers and antioxidants [159], additives [160], and volatile components of rubber in air [160, 161].

HPLC techniques are being increasingly employed as for the determination of dithiocarbamates in vulcanized rubber products [162].

Thermogravimetric analysis of individual rubber and vulcanizates have been employed in conjunction with I.R., GC or differential scanning colorimetry [163, 164].

References

1. Fishbein, L., Scand. J. Work Environmental Health 9 (Suppl. 2), 7 (1983).
2. Holmberg, B., Sjostrom, B.; Olsson, S., Investig. Rept. 19 National Board of Occupational Safety and Health, Stockholm pp. 1–127 (1977).
3. IARC. IARC Monographs on the Evaluation of the Carcinogenic Risk of Chemicals to Humans Vol. 28. The Rubber Industry. International Agency for Research in Cancer, Lyon, 452 pp. (1982).
4. Hedenstedt, A., et al., Mutat. Res. 68, 313 (1979).
5. Davis, A. J., Rubber World. 181, 29 (1980).
6. Greek, B. F., Chem. Eng. News. March 21, pp. 25–29, 36–41, 46–50 (1988).
7. Greek, B. F., Chem. Eng. News, March 31, pp. 17–24, 29–34, 38–44 (1986).
8. Dworkin, D. and Winston, J. M., Chem. Week, April 20, pp. 21–35 (1988).
9. Greek, B. F., Chem. Eng. News, May 18, pp. 29–31, 34–37, 43–49.
10. Taylor, R. and Son, P. N. Rubber Chemicals. In: Kirk-Othmer Encyclopedia of Chemical Technology, Wiley-Interscience, New York, pp. 337–364 (1977).
11. Hybart, F. J., Briscoe, G. B., and Daniel, T. J., J. Inst. Rubber Ind., 2, 190–193 (1968).
12. IARC. IARC Monographs for the Evaluation of the Carcinogenic Risk of Chemicals to Humans Vol. 16. Some Aromatic Amines and Related Nitro Compounds-Hair Dyes, Colouring Agents and Miscellaneous Industrial Chemical International Agency for Research on Cancer, Lyon, pp. 325–341 (1978).
13. Nutt, A. R. Progr. Rubber Technol., 42, 141–154 (1979).
14. U.S. Food and Drug Administration. Code of Federal Regulations 21 CFR 177.2550-177-2710.
15. Shaver, T. W., Rubber Chemicals In: Encyclopedia of Polymer Science and Technology; Vol. 12 (eds) Mark, H. F. and Gaylor, N. G., John Wiley, New York, pp. 256–290 (1970).

16. Fajen, J. M., Carson, G. A., Rounbehler, D., et al. Science, 205, 1262–1264 (1979)
17. IARC. IARC Monographs On the Evaluation of the Carcinogenic Risk of Chemicals to Humans. Vol. 17, Some N-Nitroso Compounds. International Agency for Research Cancer, Lyon (1978).
18. Ireland, C. B., Hytrek, F. P., Lasojki, B. A., Am. Ind. Hyg. Assoc. J., 41, 895–900 (1980).
19. McGlothin, J. D., Wilcox, T. C., Fajen, J. M. and Edwards, G. S. In: Choudhary, G. (ed. Chemical Hazards in the Workplace, Measurement and Control. American Chemical Society, Washington, D.C. pp. 283–299 (1981).
20. Preussmann, R., Spiegelhalder, B., and Eisenbrand, G. In: Pullman, B., TS'O, P., and Gelboin, H. (eds). Carcinogenesis: Fundamental Mechanisms and Environmental Effects, D. Reidel, Dordrecht, pp. 273–285 (1980).
21. Spiegelhalder, B., Preussmann, R. In: Bartsch, H., O'Neill, K., et al. (eds). N-Nitroso Compounds: Occurrence and Biological Effects. IARC SCI. Publ. No. 41., International Agency for Research on Cancer, Lyon (1983).
22. Yeager, F. W., Van Gulick, N. N., and Lasoski, B. A., Am. Ind. Hyg. Assoc. J., 41, 148–150 (1980).
23. Rappe, K. C., and Rydstrom, T. In: Walker, E. A., Gricute, L. Castegnaro, M. and Borzonys, M. (eds.) N-Nitroso Compounds: Analysis, Formation and Occurrence. IARC. SCI Publ. No. 31 International Agency for Research on Cancer, Lyon pp. 565–574 (1980).
24. Willoughby, B. G., Lawson, G., Rubber Chem. Technol., 54, 311–330 (1981).
25. Rappaport, S. M. and Fraser, D. A., Anal. Chem., 48, 476–481 (1976).
26. Brock, M. J., and Lonth, G. D., Anal. Chem., 27, 1575–1580 (1955).
27. Kleeman, W., and Erben, G., Rubber Chem. Technol., 37, 204–209 (1964).
28. Manik, S. P., and Banerjee, S., Rubber Chem. Technol., 43, 1311–1326 (1970).
29. Potts, K. T., Brugez, E. G., D'Amilo, J. J., and Morita, E., Rubber Chem. Technol., 45, 160–172 (1972).
30. Groves, J. S., and Small, J. M., Brit. J. Ophthalmol., 53, 683–687 (689).
31. Lieb, R. I., Sullivan, A. B., and Trivette, C. D., Jr., Rubber Chem. Technol., 43, 1188–1193 (1970).
32. Hedenstedt, A., Ramel, C., and Wachtmeister, C. A., J. Toxicol. Environ. Hlth., 8, 805–814 (1981).
33. National Toxicology Program., Toxicology and Carcinogenesis Studies of 1,3-Butadiene (Cas. No. 106-99-0) in B6C3F$_1$ Mice (Inhalation Studies) (Technical Report Series No. 288. Research Triangle Park, N.C. (1984).
34. IARC. IARC Monographs on the Evaluation of the Carcinogenic Risk of Chemicals to Humans. Vol. 39 Some Chemicals Used in Plastics and Elastomers. International Agency for Research on Cancer, Lyon, pp. 155–179; 195–226 (1986).
35. deMeester, C., Poncelet, F., Roberfroid, M., and Mercier, M. Toxicol. Lett., 6, 125–130 (1980).
36. Poncelet, F., deMeester, C., Duveger-Van Bogaert, M., et al., Arch. Toxicol., Suppl. 4, 63–66 (1980).
37. Malvoisin, E., Lhoest, G., Poncelet, F., et al., J. Chromatog., 178, 419–425 (1979).
38. Cunningham, M. J., Choy, W. M., Arce, G. T., Mutagenesis 1, 449–452 (1986).
39. Sharief, Y., Brown, A. M., Backer, L. C., Campbell, J. A., et al., Environ. Mutagenesis 8, 439–448 (1986).
40. Autrup, A., Wefald, F. C., Jeffrey, A. M., Int. J. Cancer 25, 293–300 (1980).
41. IARC. IARC Monographs on the Evaluation of the Carcinogenic Risk of Chemicals to Humans. Vol. 19. Some Monomers, Plastics and Synthetic Elastomers, and Acrolein, International Agency for Research on Cancer, Lyon, pp. 73–113 (1979).
42. Bigner, D. D., Bigner, S. H., Burger, P. C., et al., Food Chem. Toxicol. 24, 129–137 (1986).
43. Quast, J. F., Humiston, C. G., Wade, C. E., et al., Toxicologist, 1, 129 (abst. No. 467) (1981).
44. Maltoni, C., Ciliberti, A., and Carretti, D., Ann. N.Y. Acad. Sci. 381, 216–249 (1982).
45. IARC. IARC Monographs on the Evaluation of Carcinogenic Risks to Humans. Suppl. 7. Overall Evaluations of Carcinogenicity: An Updating of IARC Monographs Vol. 1–42, International Agency for Research on Cancer, Lyon (1987).
46. Milvy, P., and Wolff, M., Mutat. Res., 48, 271–278 (1977).

47. deMeester, C., Duverger-Van Bogaert, M., et al., Toxicology, 13, 7–15 (1979).
48. Cerna, M., Kocisova, J., Kodytkova, J., et al., In: Gut, I., Cikrt, M. and Plaa, G. L. (eds). Industrial and Environmental Xenobiotics, Springer-Verlag, Berlin, Heidelberg, New York, pp. 251–254 (1981).
49. Venitt, S., Bushel, C. T., and Osborne, M., Mutat. Res. 45, 283–288 (1977).
50. Leonard, A., Garny, V., Poncelet, F., and Mercier, M., Toxicol. Lett., 7, 329–334 (1981).
51. Rabello-Gay, M. N., and Ahmed, A. E., Mutat. Res., 79, 249–255 (1980).
52. Thiess, A. M., and Fleig, I., Arch. Toxicol., 41, 149–152 (1978).
53. Kawachi, T., Yahagi, T., Kada, T., et al., In: Montasano, R., Bartsch, H., and Tomatis, L. (ed). Molecular and Cellular Aspects of Carcinogen Screening Tests. IARC Sci. Publ. No. 27. International Agency for Research on Cancer, Lyon, pp. 323–330 (1980).
54. Murray, R.J., Nitschke, K. D., John, J. A., Smith, J. F., et al., Report from Dow Chemicad Co., (USA) Midland, MI (1976).
55. Murray, R. J., Nitschke, K. D., John, J. A., Smith, J. F., et al., Report from Dow Chemical Co. (USA), Midland MI, May 31 (1978).
56. Beliles, R. P., and Mueller, S. Litton Bionetics Rept. (LBI Project No. 2660) submitted to Chemical Manufacturers Assoc., November (1977).
57. National Cancer Institute. Bioassay of Styrene for Possible Carcinogenicity. Tech. Rept. Serv. No. 185. DHEW Publ. No. (NIH) 79-1741, U.S. Government Printing Office, Washington, D.C. (1979).
58. IARC. IARC Monographs on the Evaluation of Carcinogenic Risks to Humans. Vol. 36 Allyl Compounds, Aldehydes, Epoxides and Peroxides. International Agency for Research on Cancer, Lyon, pp. 245–263 (1985).
59. Vainio, H., Paakkunen, R., Ronnholm, K. et al., Scand. J. Work Environ. Hlth., 3, 147–151 (1979).
60. deMeester, C., Poncelet, F., Roberfroid, M., et al., Mutat. Res., 56, 147–152 (1977).
61. Sikov, M. R., Cannon, W. C., Carr, D. B., Miller, R. A., et al., J. Appl. Toxicol. 6, 155–164 (1986).
62. National Toxicology Program. Toxicology and Carcinogenesis Studies of Ethylene Oxide in B6C3F$_1$ Mice (Inhalation Studies) (NIH Publ. No. 86-2582) Research Triangle Park, N.C. (1986).
63. Garman, R. H., Stellings, W. M., and Maronpot, R. P., Food Chem. Toxicol. 24, 145–153 (1986).
64. IARC. IARC Monographs on the Evaluation of Carcinogenic Risks tu Humans. Suppl. 6 Genetic and Related Effects. An Updating of Selected IARC Monographs from Volume 1–42, International Agency for Research on Cancer, Lyon, pp. 301–303 (1987).
65. National Toxicology Program. Toxicology and Carcinogenesis Studies of 2-Mercaptobenzothiazole in F344/N Rats and B6C3F$_1$ Mice. NIH Publ. No. 88-2588. Research Triangle Park, N.C. May 1988.
66. IARC. IARC Monographs on the Evaluation of Carcinogenic Risk of Chemicals to Man. Vo. 12. Some Carbamates, ThioCarbamates and Carbazides, International Agency for Research on Cancer, Lyon (1976).
67. Hedenstedt, A., Rannug, U., Ramel, C., and Wachtmeister, C. A., Mutat. Res., 68, 313–325 (1979).
68. Hedenstedt, A. In: International Symposium Prevention of Occupational Cancer, Finnish Institute of Occupational Health, Helsinki (1982).
69. Korhonen, A., Hemminki, K., and Vainio, H. Scand. J. Work Environ. Hlth. (1982).
70. Korhonen, A., Hemminki, K., and Vainio, H. Teratogenesis Carcinog. Mutagenesis (1982).
71. IARC. IARC Monographs on the Evaluation of Carcinogenic Risk of Chemicals to Man. Vol. 7. Some Antithyroid and Related Substances, Nitrofurans and Industrial Chemicals. International Agency for Research on Cancer. Lyon, pp. 45–52 (1974).
72. Khera, K. S., Teratology. 7, 243–252 (1973).
73. Ruddick, J. A., and Kehra, K. S. Teratology 12, 277–281 (1975).
74. Korhonen, A., Hemminki, K., and Vainio, H. Acta Pharmacol. Toxicol. (1982).
75. Mortelmans, K., Haworth, S., Lawlor, T. et al., Environ. Mutagenesis 8 (Suppl. 7), 1–119 (1986).
76. Bempong, M. A., Mantley, R. Jepto 6, 293–302 (1985).
77. NIOSH. Current Intelligence Bulletin: Metabolic Precussors of a Known Human Carcinogen,

beta-Naphthylamine. U.S. Dept. Health, Education and Welfare, Rockville, MD pp. 1–3 (1976).

78. Kummer, R. and Tordoir, W. F., Tidjschr Geneeskd, 5., 415 (1975).
79. Innes, J. M., Ulland, B. M., Valerio, M. G., et al., J. Natl. Cancer Inst. 42, 1101–1127 (1969).
80. Veys, C. A. J. Natl. Cancer Inst. 43, 219–226 (1983).
81. Slaga, T. G., Klein-Szanto, A. J. P., Triplet, L. C., et al., Science 213, 1023–1027 (1981).
82. Fishbein, L. In: Symposium on Occupational Hazards Related to Plastics and Synthetic Elastomers, Espoo, Finland, Nov. 22–27 (1982).
83. National Toxicology Program. Carcinogenesis bioassay of di (2-ethylhexyl)-phthalate in F344 Rats and B6C3F₁ Mice (feed study). Tech. Rept. Serv. No. 217 NIH Publ. No. 82-1173. Research Triangle Park, N.C. (1982).
84. National Toxicology Program. Carcinogenesis Bioassay of Di (2-ethylhexyl)-adipate in F344 rats and B6C3f₁ Mice (feed study). Tech. Rept. Serv. No. 212. NIH Publ. No. 82-1788. Research Triangle Park, N.C. (1982).
85. IARC. IARC Monographs on the Evaluation of the Carcinogenic Risk of Chemicals to Humans Vol. 29 Some Industrial Chemicals and Dyestuffs, International Agency for Research On Cancer, Lyon, pp. 93–148, 391–398 (1982).
86. Kuna, R. A., and Kapp, R. W., Jr., Toxicol. Appl. Pharmacol. 57, 1–7 (1981).
87. IARC. IARC Monographs on the Evaluation of the Carcinogenic Risk of Chemicals to Humans. Vol. 41. Some Halogenated Hydrocarbons and Pesticide Exposures. International Agency for Research Cancer, Lyon pp. 43–85 (1986).
88. IARC. IARC Monographs on the Evaluation of the Carcinogen Risk of Chemicals to Humans. Vol. 20. Some Halogenated Hydrocarbons. International Agency for Research Cancer, Lyon, pp. 545–572 (1979).
89. Henschler, D., Elsasser, H. M., Romen, W., and Eder, E. J., Cancer Res. Clin. Oncol. 107, 149–156 (1984).
90. National Toxicology Program. Carcinogenesis Bioassay of Trichlorethylene in F344 Rats and B6C3F₁ Mice (Gavage Studies). Tech. Rept. Serv. No. 243, NIH Publ. No. 82-1799, Research Triangle Park, N.C. (1982).
91. Norpoth, K., Reisch, A., and Heinecke, A. In: Norpoth, K. H. and Garner, R. C. (eds). Short-Term Test Systems for Detecting Carcinogens. Springer-Verlag, New York pp. 31–322 (1980).
92. Barnes, J. R., and Ranta, K. E. Toxicol. Appl. Pharmacol., 23, 271–276 (1972).
93. Stula, E. F., and Krauss, W. C., Toxicol. Appl. Pharmacol., 41, 35–55 (1977).
94. Schwetz, B. A., Leong, B. K. J., and Gehring, P. G., Toxicol. Appl. Pharmacol. 28, 452–457 (1974).
95. IARC. IARC Monographs on the Evaluation of Carcinogenic Risk of Chemicals to Man. Vol. 3. Certain Polycyclic Aromatic Hydrocarbons and Heterocyclic Compounds. International Agency for Research on Cancer, Lyon (1973).
96. Rivin, D., Dangerous Prop. Ind. Mater. 5, 1–11 (1985).
97. IARC. IARC Monographs on the Evaluation of Carcinogenic Risk of Chemicals to Man. Vol. 33. Polynuclear Aromatic Compounds., Part 2, Carbon Blacks, Mineral Oils and Some Nitro Arenes. International Agency for Research on Cancer, Lyon, pp. 35–85 (1984).
98. IARC. IARC Monographs on the Evaluation of Carcinogenic Risk of Chemicals to Man. Vol. 35. Polynuclear Aromatic Compounds Part 4. Bitumens, Coal-tars and Derived Products, Shale Oils and Soots, International Agency for Research on Cancer, Lyon 39–81, 83–150 (1985).
99. IARC. IARC Monographs on the Evaluation of Carcinogenic Risk of Chemicals to Man. Vol. 3. Certain Polycyclic Aromatic Hydrocarbons and Heterocyclic Compounds. International Agency for Research on Cancer, Lyon (1973).
100. Hermann, M., Durand, J. P., Charpentier, J. M., Intl. In: Fourth International Symposium on Polynuclear Aromatic Hydrocarbons, Columbus, OH.
101. Williams, T. M., Harris, R. L., Arp, E. W., et al., Am. Ind. Hyg. Assoc. J., 41, 204–211 (1980).
102. Van Ert, M. D., Arp, E. W., Harris, R. L., et al. Am. Ind. Hyg. Assoc. J., 41, 212–219 (1980).
103. Parkes, H. G., Whittaker, B., and Willoughby, B. G., The Monitoring of the Atmospheric

Environment in U.K. Tyre Manufacturing Work Areas., British Rubber Manufacturer's Association, Ltd. Birmingham, U.K. (1975).

104. Cocheo, V., Bellomo, M. L., and Bombi, G. G., Am. Ind. Hyg. Assoc. J. 44, 521–527 (1983).
105. McGlothin, J. D., Wilcox, T. C., Fajen, J. M., and Edwards, G. S. In: Choudharry, G., (eds). Chemical Hazards in the Workplace Measurement and Control. ACS Syposium Series 149. American Chemical Society. Washington, D.C., pp. 283–299 (1981).
106. McMichael, A. J., Spirtas, R., and Kupper, L. L., J. Occup. Med., 16, 458–464 (1974).
107. McMichael, A. J., Spirtas, R., Kupper, L. L., and Gamble, J. F., J. Occup. Med., 17, 234–239 (1975).
108. Hakama, M., and Kilpikari, I., In: Occupational Cancer and Carcinogenesis. (eds) Vainio, H., Sorsha, M., and Hemminki, K., Hemisphere Publ. Co., Washington, D.C., pp. 1211–1218 (1979).
109. Checkoway, H., and Williams, T. M., Am. Ind. Hyg. Assoc. J., 43, 164–169 (1982).
110. Mancuso, J. F., In: Levinson, C., (ed), New Multinational Health Hazards, (ed), ICF Geneva, pp. 80–136 (1975).
111. Fox, A. J., Lindars, D. C., and Owen, R., Brit. J. Ind. Med., 31, 140–151 (1974).
112. Fox, A. J., and Collier, Brit. J. Ind. Med. 33, 249–264 (1976).
113. Checkoway, H., Wilcosky, T., Wolf, P. and Tyroler, H. Am. J. Ind. Med., 5, 239–249 (1984).
114. Wilcosky, T., Checkoway, H., Marshall, E-6, and Tyroler, H. A. Am. Ind. Hyg. Assoc. 45, 808–811 (1984).
115. Kilpikari, I., Acta Universites Tamerensis. Ser. A. Vol. 138, pp. 7–71 (1982).
116. Wolf, P. H., Andjeckovich, D., Smith, A., and Tyroler, H., J. Occup. Med., 23, 103–108.
117. Delzell, E., and Monson, R. R., J. Occup. Med., 24, 767–770 (1982).
118. Wang, H. W., You, X. J., Qu, Y. H., et al., Cancer Res., 44, 3101–3105 (1984).
119. Monson, R. R., and Nakano, K. K., Am. J. Epidemiol., 103, 284–296 (1976).
120. Monson, R. R., and Nakano, K. K., Am. J. Epidemiol., 103, 297–303 (1976).
121. Monson, R. R. and Fine, L. J., J. Natl. Cancer Inst., 61, 1047–1053 (1978).
122. McMichael, A. J., Andjelkovich, D. A., and Tyroler, H. A., Ann. N.Y. Acad. Sci., 271, 125–137 (1976).
123. British Rubber Manufacturer's Assoc. (BRMA) Health Research Project Report, BRMA, Birmingham U.K. (1976).
124. Case, R. A. M., and Hosker, M. E., Brit. J. Prev. Soc. Med., 8, 39–50 (1954).
125. Liira, J., Riihimaki, V., and Pfaffli, P., Int. Arch. Occup. Environ. Hlth., 60, 195–200 (1988).
126. Goldman, R. H., Lancet 2, 744–745 (1979).
127. Saida, K., Mendell, J. R., and Weiss, H. S., J. Neuropathol. Exp. Neurol. 35, 207–225 (1976).
128. Ralston, W. H., Hildebrand, R. L., Uddin, D. E., et al., Toxicol. Appl. Pharmacol. 81, 313–327 (1985).
129. WHO, Environmental Health Criteria No. 28 Acrylonitrile, World Health Organization, Geneva (1983).
130. WHO, Environmental Health Criteria No. 26. Styrene, World Health Organization, Geneva (1983).
131. Lee, C. L., and Peters, P. J., Environ. Hlth. Persp., 17, 35–43 (1976).
132. Snyder, R. H., Vincent, V. R., and Querry, F. C., National Tire Disposal Symposium, Washington, D. C., June, (1977).
133. Paul, J., Rubber. In: Grayson, M. (ed). Kirk-Othmer Encyclopedia of Chemical Technology, 3rd ed., Vol. 19. John Wiley & Sons, New York, pp. 1002–1010.
134. Morton, M., Rubber Technology, 2nd ed. Van Nostrand-Reinhold, New York (1973).
135. Bebb, R. L., Environ. Hlth. Persp., 17, 95–101 (1976).
136. Barnhardt, R. A. In: Standen, A. (ed). Kirk-Othmer Encyclopedia of Chemical Technology, 2nd ed., Vol. 17. John Wiley & Sons, New York, pp. 581–590.
137. Fraser, D. A., and Rappaport, S., Environ. Hlth Persp., 17, 45–53 (1976).
138. Penn, W. S., Plastics 15, 18 (1950).
139. Bullard, H. L., Rubber Chem. Technol., 37, 210–215 (1964).
140. Bullard, H. L., Rubber Chem. Technol., 38, 134–136 (1965).
141. Anon, Pesticide & Toxic Chem. News. 16, 11 (1988).
142. Gaeta, L. J., Schleuter, E. W., and Altenau, A. G., Rubber Age, 101, 47–50 (1967).

143. Tsuchii, A., Suzuki, T., and Takahara, Y., Agric. Biol. Chem., 52, 1217–1222 (1978).
144. Williams, G. R., Int. Biodeterior., 20, 173–175 (1984).
145. Verschueren, K., Handbook of Environmental Data on Organic Chemicals. Van Nostrand Reinhold, New York, (1977).
146. Patil, S. S., and Shinde, B. M., Environ Sci. Technol., 22, 1160–1165 (1988).
147. Davis, B. D., and Morgan, R. C., In: Morris, C. R. and Cabral, J. R. P. (eds) Hexachlorobenzene: Proceedings of an International Symposium, International Agency for Research on Cancer, Lyon, pp. 23–30.
148. Sourisseau, R., and Mueller, W., Plaste. Kautsch., 34, 447–448 (1987)
149. Alekseeva, K. V., and Solomatina, L. S., Kauch. Rezina., 8, 54–55 (1978) [Chem. Abstr., 89, 147906, (1978)].
150. Alekseeva, K. V., J. Anal. Appl. Pyrolysis., 2, 19–34 (1980).
151. Hausler, K. G., Schroeder, E., and Huster, P., J. Anal. Appl. Pyrolysis, 2, 109–121 (1980).
152. Hausler, K. G., Schroeder, E., and Huster, P., Plaste. Kautsch., 27, 548–552 (1980).
153. Krishen, A., Astm. Spec. Tech. Publ. 553, 74–86 (1974).
154. Hu, J. C. A., Anal. Chem., 49, 537–540 (1977).
155. Fritsch, H., Labor Praxis., 11, 1092–1098 (1987).
156. Lapshova, A. Y., Ivanova, M. P., and Chechetkino, L. N., Zh. Anal. Khim., 42, 1114–1118 (1987).
157. Deome, A. J., Bulpett, D. A., and Kulig, C. J., Sagamore Army Mat. Res. Conf. Proc. 1985 32nd Elastomers. Technol. pp. 307–316 (1987) [Chem. Abstr. 107, 41285 (1987)].
158. Kochel, I., and Jaroszynskya, D., Polimery (Warsaw), 23, 218–220 (1974).
159. Johnson, El., Gloor, A., and Majors, R. E., J. Chromatog., 149, 571–578 (1978).
160. Drugov, Y. S., and Muraveva, G. V., Zh. Anal. Khim., 34, 2252–2259 (1979).
161. Casper, R. H., Annual Mtg. Proc. Int. Inst. Synth. Rubber Prod., 18 (1979).
162. Kaniwa, M. J., J. Chromatogr., 405, 263–271 (1987).
163. Rugo, G., Barbattini, D., Cappellazo, G., et al. Ind. Gomma. 24, 31–34 (1980).
164. Sirkar, A. K., and Lamond, T. G., Rubber Chem. Technol., 51, 647–654 (1978).

Organolead Compounds

R. J. A. Van Cleuvenbergen and F. C. Adams

University of Antwerp (U.I.A.) Universiteitsplein 1, B-2610 Wilrijk (Belgium)

Summary

Since the introduction of tetraalkyllead compounds as antiknock agents in gasoline, a major part of the lead burden of the biosphere appears to be related to the commercial use of organolead products. Although historically the anthropogenic lead burden has been almost entirely attributed to the inorganic compounds, the specific hazard from organic lead cannot be considered negligible. This has led to a growing interest in the environmental pathways of these organometallic compounds. Recently, methods for sensitive speciation have clearly provided a key to elucidate further the biogeochemical cycle of organic lead.

The present survey of the literature was set up against this background to ascertain in a critical way the progress made in evaluating the precise nature and extent of the burden caused by organic lead. After a short review on the properties and synthesis of organic lead, the evolution of the analytical techniques for determination of alkyllead species is described. Next, after having revised anthropogenic and possible natural sources of pollution, the article focuses on the concentrations of the organolead species in the environment, updating the data set and discussing in detail the relevant implications for the biogeochemical cycle of organolead. An up-to-date status of the sink processes and health aspects completes this survey on occurrence and fate of organolead in the biosphere.

Introduction

Lead, being one of the most permanent of the metals, has since long been recognized as an ubiquitous pollutant in the environment. The corrosion-resistant and low melting metal already attracted mankind in prehistory, the earliest metallurgical applications apparently dating back as far as 7000–5000 B.C. [1]. During the Roman Empire the gradually increasing production of lead had reached about 4 kg capita^{-1} yr^{-1} [2], most of it being used in the manufacture of water pipes and in cooking utensils. This led to the suggestion that the fall of the Roman culture was largely due to this extensive domestic use and the resulting contamination of food and drink [3]. After a temporary decline in lead production, a continuous growth can be noticed from the Middle Ages on, with particularly the industrial revolution (1750) implying a revival of large-scale lead smelting. The discovery in 1922 of the antiknock properties of lead compounds in the gasoline of motor vehicles [4] opened the door to a new flourishing of lead mining and refining. Today, the total world production of lead can be estimated at about 3.5 million ton annually [5], the automobile and construction industries being the main consumers. The manufacture of lead batteries accounts for approximately 50 % of the consumption, whereas lead pigments and chemicals further contribute significantly [6].

Organolead, on the other hand, has a much more recent history. The first reported synthesis of a compound containing a lead-carbon bond appears to be that of Löwig [7] who in 1852 produced impure hexaethyldilead from a sodium lead alloy and ethyl iodide. Exposed to air, it was transformed to a white crystalline compound which proved to be triethyllead carbonate. In 1859, Buckton prepared tetraethyllead and showed that it could be converted into triethyllead salt by the action of acids [8]. The synthesis of organolead compounds from Grignard reagents by Pfeiffer and Truskier [9] provided a suitable strategy for the systematic preparation of tetraalkyl- or tetra-aryllead compounds containing equal or different organic radicals, and the corresponding salts. Soon, around 1920, the isolation of tri- and divalent organic lead compounds followed. No practical application had been found, nor perhaps thought of, for any of the organolead products until the discovery in the research laboratories of General Motors that tetraethyllead efficiently acts as an antiknock agent in gasoline engines [4]. This breakthrough gave a powerful impetus to organolead chemistry: a real antiknock industry was developed, and many aspects of the field that had been investigated only superficially, like the toxicity [10], were further elucidated.

Since the introduction of tetraalkyllead compounds as additive in gasoline, a major part of the lead burden of the biosphere appears to be related to the commercial use of organolead products. The significance of this anthropogenic pollution source became impressively reflected by numerous total lead records [e.g. 11–17]. This fact, in addition to the poisoning effect of lead compounds on the catalysts which are currently being implemented to purify the car exhaust gases, resulted in organic lead featuring in the on-going debate on the reduction of vehicular emissions [18]. In 1975, the United States and Japan became the first countries to gradually switch to unleaded gasoline and catalytic converters. Now, about a decade later, the beneficial effect of this controversial decision on the environmental lead level seems evident, as illustrated in Figure 1. Western Europe has followed the trend and will enforce emission limits broadly in line with the U.S. standards between 1989 and 1993; initially a progressive reduction of the lead content in gasoline is being carried through, phasing down the level to or below 0.15 g per litre [19].

The present survey of the literature was set up against this background to ascertain in a critical way the precise nature and extent of the burden caused by organic lead. Although the specific form of the compounds emitted into the biosphere due to the widespread use of organolead is largely inorganic, reflecting the relative instability of organolead species, the hazard from organic lead cannot be considered negligible. This has led to a growing interest in the environmental pathways of these organometallic compounds. The recent optimization of sensitive speciation methodologies is clearly providing a key to elucidate further the biogeochemical cycle of organic lead. The present review will attempt to present an up-to-date status of our knowledge

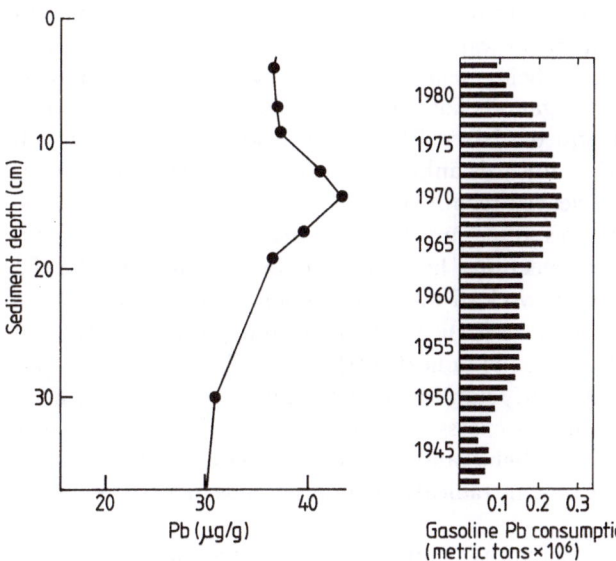

Fig. 1. (left) Profile of total lead concentrations in sediment plotted against depth; the samples were taken from the Mississippi Delta, United States, in 1983 at a water depth of 69 m (right) Annual consumption of lead in gasoline additives in the United States from 1942 to 1983. The dates indicated in the right part correspond with the dates which were calculated for selected sediment depths from ^{210}Pb geochronologies.
(Modified from reference 15)

of organolead in the biosphere, supplementing existing surveys of this topic [5, 20, 21]. The environmental aspects associated with inorganic lead are treated in another part of this handbook.

Properties of Organolead Compounds

Organometallic compounds behave fundamentally distinct both in chemical and biological properties from ionic compounds of the same metal. The availability to organisms and the toxicity of a heavy metal may differ considerably from one chemical species of the metal to another. Despite the huge diversity of synthetic species, the environmental chemistry of organolead is dominated by a relatively small number of products. It is determined almost exclusively by 5 tetraalkyllead species, being tetramethyllead, tetraethyllead and the mixed methyl-ethyl species, and further the 9 organolead salts which can possibly result from these 5 species through successive dealkylation. Table 1 shows a schematic survey of these compounds, on which this chapter is mainly focusing.

Organolead derivatives can be divided into two main groups. A first class consists of those compounds in which the lead atom is bonded exclusively to carbon or to another lead atom. With tetraorganolead species being the most typical representatives, this class further includes hexaorganodilead and diorganolead species, which are at present merely interesting on theoretical grounds. The low molecular weight tetraalkyllead compounds are clear, colorless liquids, whereas the high molecular weight derivatives are solids. Taking into account the small difference in the electronegativities of lead and carbon, the lead-to-carbon bond in tetravalent lead compounds, to which the lead atom contributes a sp^3 hybrid orbital, exhibits a high degree of covalent character. This is reflected by the physico-chemical properties which characterize the compounds [22–27]. Tetraalkylleads are soluble in the common organic solvents, such as ether, benzene, chloroform, or absolute ethanol. The symmetrical tetraarylleads, however, appear to be almost insoluble in alcohols or ethers. Generally R_4Pb compounds are quite insoluble in water and, chemically, fairly stable to water and air. Tetraorganolead species, being relatively unreactive compared to other organolead derivatives, will resist organometallic reactions like addition to a carbonyl group, as well as reactions with aqueous bases. With aqueous strong acids, a reaction takes place at moderate rate, while towards anhydrous acids or halogens the carbon-lead bond is quite sensitive and cleaved readily and completely. Under more drastic conditions (at elevated temperatures, often in solution in organic solvents) the compounds undergo a wide variety of reactions, both electrophilic and nucleophilic. To produce selective cleavage of one or two lead-to-carbon bonds, a strict control of the reaction stoichiometry and temperature is then usually a prerequisite. Pyrolysis of organolead compounds gives free organic radicals [28]. Although PbR_4 compounds (R = organic radical) are generally fairly stable to light and can be stored for extended periods in brown glass bottles without excessive decomposition, if air is rigorously excluded, irradiation with ultraviolet light causes decomposition of tetraorganolead compounds to lead metal and free organic radicals; photolysis tends to produce similar results as pyrolysis.

A second class of organolead derivatives contains those compounds build up by

Table 1. Schematic survey of environmentally important organolead compounds.

TAL →	$CH_3-Pb(CH_3)(CH_3)-CH_3$ tetramethyllead	$CH_3-Pb(C_2H_5)(CH_3)-CH_3$ trimethylethyllead	$CH_3-Pb(C_2H_5)(C_2H_5)-CH_3$ dimethyldiethyllead	$CH_3-Pb(C_2H_5)(C_2H_5)-C_2H_5$ methyltriethyllead	$C_2H_5-Pb(C_2H_5)(C_2H_5)-C_2H_5$ tetraethyllead
triAL →	$CH_3-Pb^+(CH_3)-CH_3$ trimethyllead	$CH_3-Pb^+(C_2H_5)-CH_3$ dimethylethyllead	$CH_3-Pb^+(C_2H_5)-C_2H_5$ methyldiethyllead	$C_2H_5-Pb^+(C_2H_5)-C_2H_5$ triethyllead	
diAL →	$CH_3-Pb^{++}-CH_3$ dimethyllead	$CH_3-Pb^{++}-C_2H_5$ methylethyllead	$C_2H_5-Pb^{++}-C_2H_5$ diethyllead		
[monoAL] →	CH_3-Pb^{+++} monomethyllead	$C_2H_5-Pb^{+++}$ monoethyllead			
Pb^{++} →					

one or more lead-carbon bonds, but in addition by at least one bond to an atom other than carbon or lead. Most of the known compounds which form the second class are of the types R_3PbX, R_2PbX_2 and $RPbX_3$, where X is an anionic group, usually a halogen or a carboxylate group. Through dipole moment measurements on several selected tri- and diorganolead halides, it has been demonstrated that the lead-halogen bond in these compounds possesses a high degree of ionic character, approaching that of the lead-halogen bonds in the divalent lead halides [29]. Although the lead compounds R_3PbX, R_2PbX_2 and $RPbX_3$ are commonly called 'organolead salts' it should be emphasized that they do not have an ionic structure, but are in fact coordination compounds in which the metal-halogen bond is a more or less polarized covalent bond. Consequently the solubility of e.g. R_3PbX species depends both on the radical R and the anionic group X; the solubilities in water diminish with increasing size of the radical R, undoubtedly due to decreasing solvatation, whereas the solubilities will also vary considerably between halides and compounds containing an organic radical X.

The tri- and diorganolead compounds are clearly less stable than tetraorganolead species. They are assumed to undergo a slow decomposition, even at room temperature, via a disproportionation reaction [26]:

$$2Rb_3PbX \rightarrow R_2PbX_2 + R_4Pb$$
$$2R_2PbX_2 \rightarrow R_3PbX + [RPbX_3] \rightarrow RX + PbX_2 .$$

The first reaction is reversible, the second is irreversible because of the instability of most $RPbX_3$ species. Because of the relative lability of alkyllead halides, any interpretation of the mechanism of reactions involving these salts should consider the possibility of a simple thermal disproportionation. The high degree of ionic character of the organolead halides is clearly reflected in the large number of metathesis reactions they undergo, analogous to those of the divalent lead salts. Among these, the redistribution reactions between organolead compounds, discovered by Calingaert [30, 31], should explicitly be mentioned because of their remarkable nature: firstly no ionizing solvent is required, but only some catalyst, and secondly the randomness of the re-

Table 2. Physical properties of some selected organolead compounds [21, 26]

Compound	Boiling point (°C)	Vapor pressure at 20 °C (mmHg)	Melting point (°C)	Dipole moment (Debyes)	Water solubility (mg Pb l^{-1})
tetramethyllead	110	26	− 27	0	15
trimethylethyllead	130	7.3			
tetraethyllead	202*	0.26	−137	0	≦0.1
trimethyllead chloride			190*	4.47	
triethyllead chloride			166*	4.39	20000
diethyllead dichloride				4.70	50000

* : decomposes

actions leads to a product composition obeying the laws of probability. Compounds of the types R_2PbX_2, and particularly $RPbX_3$ [32], generally tend to be less stable than the R_3PbX species. Although the majority of organolead salts are stable to hydrolysis, as evidenced by their facile preparation in aqueous media, this resistance appears to become weaker as the number of alkyl groups decreases. Practically, the organolead salts undergo similar reactions as the tetraorganolead products.

Because a detailed survey of the physico-chemical properties would be out of the scope of this chapter we refer to the outstanding compilation by Shapiro and Frey [26] for more details. Some important physical properties of those organolead species which are frequently detected in the environment are summarized in Table 2.

Synthesis of Organolead Compounds

As the general synthetic chemistry of organolead has not proceeded markedly since it was extensively reviewed elsewhere [26], only a brief summary of the frequently applied strategies in the preparation of organolead compounds will be presented here. Figure 2 shows a scheme of the pathways commonly employed. The synthesis of symmetrical tetraorganolead products is relatively straightforward. On a laboratory scale, generally reactions of lead halides with organolithium and Grignard reagents are preferred, proceeding through intermediate formation of R_2Pb and R_6Pb_2 species. Other organometallic reagents like R_2Zn, R_3Al and R_3B have also been used, but appear to be less versatile or efficient for lead alkylation. A second useful route is the reaction of lead alloys with an organic halide, which enjoys commercial significance, being still the major process for the manufacture of tetraalkyllead antiknock agents. Industrially, the reaction is carried out with sodium-lead alloy and an equimolar amount of ethyl or methyl chloride in a pressure autoclave, in the presence of a catalyst. Three-fourths of the lead values are converted back to lead metal and must be recycled; the final yield amounts to about 85%. Additionally, reactions for the synthesis of tetraorganolead starting from lead metal have been optimized, adding to it an alkyl halide, occasionally in combination with another reactive metal or organometallic reagent. Finally, many efforts have been devoted to the industrial realization of an electrolytic synthesis of the antiknock agents. The reaction of a lead anode with a complex organometallic electrolyte has been extensively studied and has led to several interesting laboratory applications [26]. Recently, the electrochemical reduction of an alkyl halide at a sacrificial lead cathode enjoyed renewed attention for the preparation of radiolabeled alkyllead standards [33].

For the preparation of unsymmetrical tetraorganolead compounds only one method is generally suitable; it is the reaction of a triorganolead halide or diorganolead dihalide with a Grignard or organolithium reagent. By treating the resultant reaction mixture with halogen at $-78\ °C$, an unsymmetrical triorganolead halide is produced in situ, which in turn can be treated with a different Grignard or organolithium reagent. Repetition of this cycle finally yields a tetraorganolead derivative containing four distinct alkyl groups. The cleavage of the organic group by halogen is not random, shorter chain alkyl groups being cleaved more easily than longer chain groups and aryl preferentially to alkyl groups. Other organometallic reagents than Grignard or organolithium may catalyze a redistribution reaction in which a mixture of all possible compounds is formed.

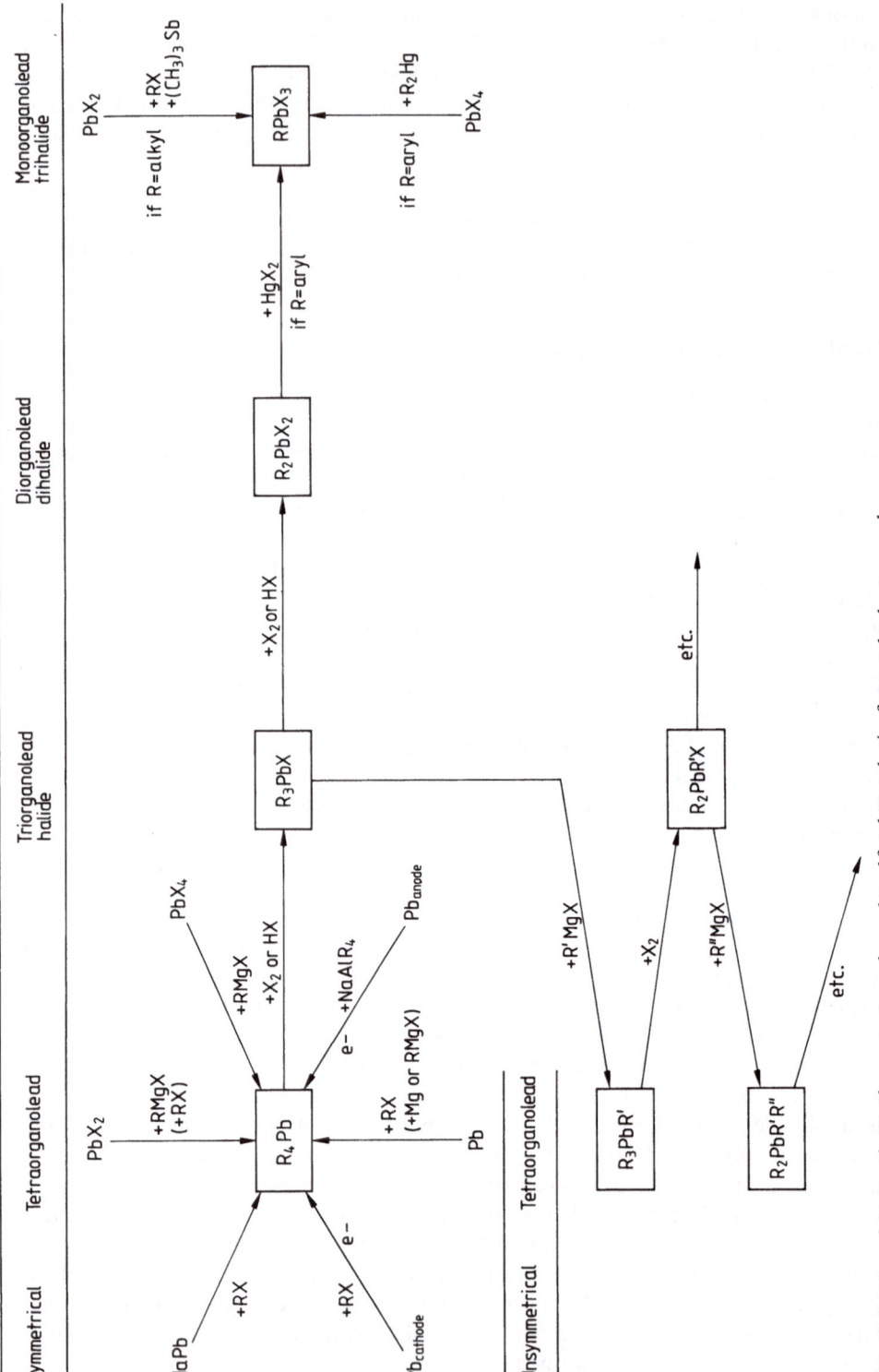

Fig. 2. Scheme of the reaction pathways commonly employed for the synthesis of organolead compounds

Numerous methods of synthesis are known for R_3PbX and R_2PbX_2 compounds, most of them involving the use of another organolead compound like R_4Pb as starting material and a halogen or halogen acid as reagent. The selectivity largely depends on the choice of the reaction parameters. Presumably due to their instability, the preparation of monoorganolead salts is more challenging, particularly that of monoalkyllead halides. To our knowledge, only one report describes the synthesis of monoalkyllead triiodides [34]; surprisingly, it was found that the compounds were stable at room temperature towards water, acids and bases.

Analytical Techniques for the Determination of Organolead Compounds

Starting from the seventies, mounting concern about the contribution of organolead to the lead burden of the biosphere has stimulated markedly the development of a sensitive analytical methodology to monitor these pollutants. During the last decade, considerable work has been done to identify and quantify alkyllead compounds in environmental matrices, in an effort to establish a quantitative model of their environmental pathways. Many of the earlier methods were suffering from two general drawbacks: a rather poor specificity, largely due to the inability to discriminate unambiguously between inorganic lead salts and (some of) the organic lead species, and a limited sensitivity, allowing measurements only in highly contaminated areas. Particularly during the evaluation of accidental local pollution, like that caused by the Cavtat incident [35, 36] the urgent need for a reliable detection of the species and their degradation products was emphasized. A major step forward has been realized by the optimization and application of hyphenated techniques for organolead analysis. The combination gas chromatography-atomic absorption spectrometry has now universally proven to be a promising key to open the remaining doors to a profound knowledge of the environmental occurrence and fate of organolead compounds.

Gaseous Fraction of the Air

For the separation between organic and inorganic lead, most of the procedures commonly used for the determination of traces of organolead compounds in the atmosphere rely on filtration. The particulate matter, assumed to contain most, if not all, of the inorganic lead, remains quantitatively on the filter while the organic lead vapors pass on for concentration and subsequent analysis. The original methods which measure the complete nonfilterable lead fraction are now generally rejected in favour of methods which are based on a species-specific determination.

Among the former methods the first attempts to estimate organic lead concentrations included a collection step with iodine or iodine monochloride [37–43]. The final detection of lead, which was often carried out colorimetrically did not rigorously exclude possible interference of inorganic lead passed through the filter. These techniques, therefore, became gradually criticized [44–49]. The selectivity of these wet chemical methods towards alkyllead species, though increased by connecting a particle filter in front of the vapor-collecting medium in the sampling train, could eventually be considered complete after the introduction of masking agents for inorganic lead in the analysis stage. The procedure of Hancock and Slater [50], who eliminated interference by chelating Pb^{++} with EDTA during the extraction of the dialkyllead species as dithizonates from the ICl scrubber solution and measured then lead via atomic absorption spectrometry after back-extraction in acid medium is still being used. It is considered a standard wet chemical method for alkyllead analysis, and has been followed or modified in several studies afterwards [51–53]. Recently, a simplified ICl method has been reported [54], introducing the annular denuder diffusion technique to avoid interference of lead particles, thus overcoming the tedious sample treatment procedures associated with the iodine monochloride based analysis of organolead.

A major drawback of all the wet chemical methods, however, remains the lack of specificity towards a particular organolead compound or class of compounds: the complete nonfilterable fraction, which may contain tetraalkyllead vapors as well as other organolead species with a sufficient volatility like trialkyllead vapor (or even inorganic lead particles escaping trapping by the prefilter in early procedures), will result in a single combined lead signal. Some relevant analytical data on these wet chemical methods for "total" nonfilterable organic lead can be compiled from the original papers. Although several definitions of criteria like the detection limit have been employed, it appears that the characteristics of the most common procedures lie close together. The minimum amount of lead detectable is generally 5–10 ng (for a comparison with other methods, see Table 3). Detection limits range between 40 and ca. 0.25 ng Pb m^{-3} for sampling periods from 1 to 48 hours. The overall precision has been estimated to be of the order of 10%.

Organic lead compounds may also be collected quantitatively by adsorption on activated carbon [55–58] and determined subsequently by a wet chemical method. The high lead blanks of the collection material, and the difficulties in desorbing the alkyllead, have so far limited a widespread application of this principle. Although interference of inorganic lead can be avoided by using an organic solvent to extract the alkyllead from the activated carbon, speciation of alkyllead compounds collected on carbon has as yet not been optimized, probably due to decomposition of part of the sampled organolead to inorganic lead [58].

The inability to obtain species-specific information using these wet chemical methods has stimulated the analysts to optimize sensitive instrumental techniques for the speciation of organolead. An enrichment technique, however, remains generally necessary to detect the trace concentrations at which the pollutants are present in the biosphere. Both cryocondensation and adsorption on porous polymers or on activated carbon have been extensively applied to realize the enrichment. Cryotrapping, being a universal method for trapping of trace contaminants, suffers from the major drawback of water clogging of the trap at the low temperatures used. As chemical drying agents tend to remove part of the organolead from the air stream, a cryogenic pretrap has been employed instead to reduce the problem [59, 60]. This led to a discussion about whether or not the cold pretrap also removes tetraalkyllead under realistic conditions [61, 62]. In most of the studies relying on cryotrapping the problem appears to have been circumvented, through adapting either the sampling parameters or the desorption procedure to take into account the traces of water sampled [63–71]. Due to the complex nature of a cryogenic trapping procedure as schematized in Figure 3, it has been found very difficult to assure invariably complete recovery of the analytes [72] although the validity of the methods seems well documented. Moreover,

Table 3. Analytical techniques for the determination of organolead compounds in the gaseous fraction of the air.

Authors	Procedure		Detection limit (ng Pb)*			Typical sample		Ref.
	Pretreatment	Detection	TML	TEL	org Pb	Volume (m³)	Flow (m³ h⁻¹)	
Hancock et al.	wet chemical (ICl)	GFAAS			11	0.030–0.090	0.180	50
Birch et al.	wet chemical (ICl)	GFAAS			ca. 4	5	0.060–0.200	51
De Jonghe et al.	wet chemical (ICl)	GFAAS			8	1	0.180	52
Røyset et al.	carbon-acid digestion	GFAAS			2	0.005	0.00001	58
Rohbock et al.	cryotrapping	GFAAS			0.2	0.200	0.060–0.120	59, 62
Harrison et al.	cryotrapping	FAAS			0.2	0.014–0.060	0.100	64
Jiang et al.	cryotrapping-acid digestion	GFAAS			14	0.400	0.360	71
Chau et al.	cryotrapping	GC-FAAS	80	80		0.010	0.008	66
Coker	adsorption	GC-FAAS	10	15			0.015	74
Radziuk et al.	cryotrapping	GC-GFAAS	0.04	0.04		0.002–0.075	0.004	60
Robinson et al.	cryotrapping	GC-GFAAS	0.1	0.1		1		67
De Jonghe et al.	cryotrapping	GC-QFAAS	0.04	0.09		0.400	0.360	69
Chau et al.	cryotrapping	GC-QFAAS	0.1	0.1			0.008	76
Hewitt et al.	adsorption	GC-QFAAS	0.02	0.03		0.080	0.003–0.025	75
Cantuti et al.	adsorption	GC-ECD		1500		0.020	0.090	73
Laveskog	cryotrapping	GC-MS	0.01	0.01		0.001		63
Nielsen et al.	adsorption-cryotrapping	GC-MS	0.06	0.01		0.015–0.200	0.003–0.007	70
Reamer et al.	cryotrapping	GC-MWPD	0.3	0.3		0.600	0.300	68

* TML = tetramethyllead, TEL = tetraethyllead, org Pb = total organic lead

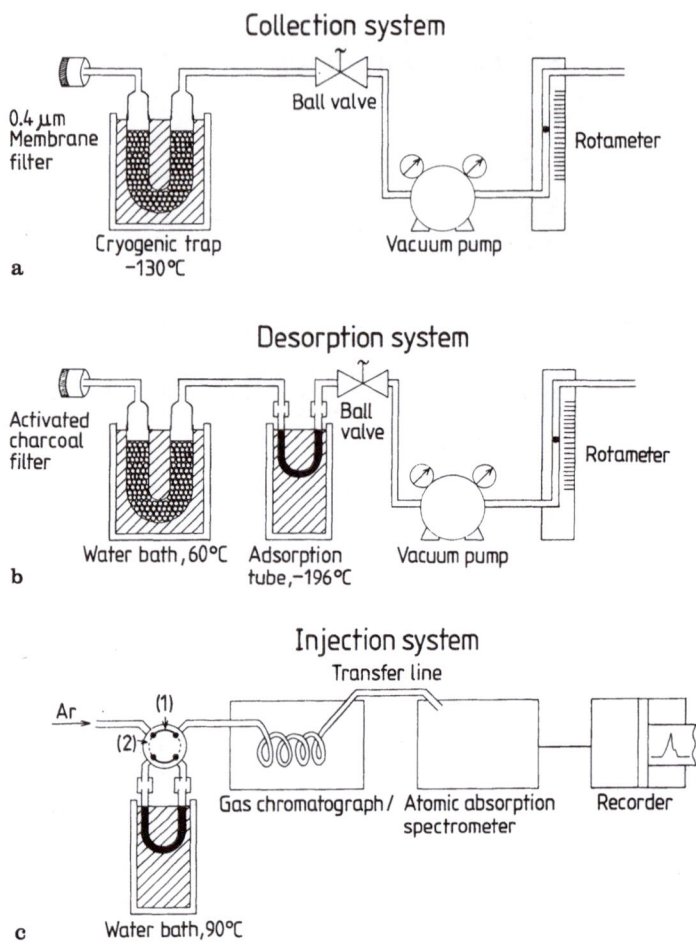

Fig. 3. Typical experimental set-up for the determination of gaseous tetraalkyllead in the air with enrichment by cryotrapping. (From reference 69)

practical disadvantages are associated with the application of this concentration technique in remote areas, and preclude its use in semi-automated sampling.

The other major enrichment technique, adsorption on porous polymers [70, 73–75], is less popular but has enjoyed continuing interest. The approach which has recently been chosen to handle the often overlooked risk of breakdown of TAL during sampling [70] is the removal of the oxidant ozone from the air stream prior to collection [75]. To ensure a reasonable sensitivity, trapping of the desorbing compounds on a cold U-tube prior to injection into the GC-AAS system remains unavoidable. Summarizing the enrichment techniques, none of both has so far proven to be an ideal one. Interlaboratory comparisons might reveal in the future the most accurate and convenient way of sampling gaseous atmospheric organolead.

Despite the complexity of sampling, results of organolead measurements based on instrumental techniques are broadly consistent, and can be divided into two groups, depending on the detection which was carried out. In particular a number of earlier instrumental procedures [59, 64, 65, 71, 74] still measure "total" nonfilterable organic lead, without attempting speciation of individual TAL compounds and other organolead products. Their concentration data may be too low to represent total organic lead, owing to the presence of part of the non-TAL species on the particulate matter, but also too high to represent only TAL. The instrumental technique applied was in most of these procedures direct AAS. The species-specific measurement of airborne TAL compounds, on the other hand, implies the use of chromatographic separation techniques. Owing to the favourable vapor pressures [26], gas-liquid chromatography has invariably been considered to give suitable separation of tetraalkyllead species. It has been combined with electron-capture detection [73], mass-spectrometry [63, 70], flame atomic absorption [66, 74], quartz [76] or ceramic [75] furnace atomic absorption, graphite furnace atomic absorption [60, 67, 69] or microwave plasma detection [68] for speciation of aiborne organolead. Not surprisingly, AAS is the most widely used, being a commonly available, highly sensitive and lead-specific instrument. Furnace atomization generally allows the better detection limits.

A thorough review on the various analytical procedures for the determination of alkyllead in the air has been published by De Jonghe and Adams [49]. In Table 3, a survey is presented of the characteristics of the most popular instrumental methods, including a comparison with other methods.

Recent investigations by Hewitt and Harrison [77–80] have thrown new light on the evaluation of the analytical procedures for airborne organolead. For the first time, indirect evidence about the presence of vapor-phase organolead other than TAL [81] was confirmed by direct measurement of gaseous trialkyllead, via collection in gas bubblers containing water, and subsequent extraction and quantification. It was claimed that the fraction of this volatile trialkyllead was significant, accounting for more than 10% of the total alkyllead. Although the possible influence of TAL on the measurements may need further verification, these studies stress once again that only speciation can unambiguously elucidate the fate and occurrence of organolead in the air.

Water Samples

As the presence of tetraalkyllead species in water is only transient due to the limited solubility particular attention has recently been focused on the determination of the ionic tri- and dialkyllead compounds in the aqueous environment. Depending on the methodology, the TAL species are first removed from a sample, e.g. by solvent extraction, or speciated simultaneously with the degradation products. Table 4 summarizes the characteristics of the most representative procedures. The separate analysis for tetraalkyllead after its enrichment by solvent extraction (usually with hexane) can be carried out in a species-specific way using common techniques based on a gas chromatographic separation [82–85]. Alternatively, thermal desorption followed by cold trapping has been employed to preconcentrate the compounds [86]. A direct sample injection method by using the GC-AAS system, has also been developed [87]. The combination GC-AAS appears to be the most versatile, sensitive and specific instrumental technique for the purpose. In the sample treatment procedure, however, care should be taken to check whether the extracted lead only represents TAL,

Table 4. Analytical techniques for the determination of organolead compounds in water samples

Authors	Procedure	Detection limit (ngPbl⁻¹) per species			Ref.
		TAL	TriAL	DiAL	
Potter et al.	extraction-GC-ECD	2000			82
Noden	extraction-MS	4000—8000			84
	head space analysis-MS	200—1000			
Cruz et al.	extraction-GFAAS	1			86
	thermal desorption-GC-GFAAS	500			
De Jonghe et al.	salting out-extraction-GFAAS		20		92
Blaszkewicz et al.	adsorption-HPLC-PAR detection		15—20		96
Chau et al.	extraction-butylation-GC-QFAAS	25—100	25—100	25—100	83, 100
Van Cleuvenbergen et al.	extraction-butylation-GC-QFAAS	ca. 20	0.1	0.2	85, 104
Radojevic et al.	extraction-propylation-GC-QFAAS	0.2—0.3	0.3—0.6	0.4—1.0	103
Baussand et al.	hydride-cold trapping-QFAAS	100—500	200—1000	ca. 1000	106, 107
D'Ulivo et al.	hydride-atomic fluorescence		3	3—5	108
Rapsomanikis et al.	in situ ethylation-cold trapping -QFAAS (only methyllead)		0.18	0.21	109
Hodges et al.	DPASV		20	20	110
Colombini et al.	extraction-DPASV		ca. 300	ca. 300	111, 112

or includes another fraction. Moreover, sample storage may drastically affect the reliability of the results, as TAL species are well known to be subject to adsorption on particular types of recipients [88]. For routine pollution monitoring where speciation is not required the final determination can be carried out with graphite furnace AAS after digestion [89, 90].

The isolation of tri- and dialkyllead species from an aqueous sample has been a more challenging task to the analyst, as their ionic characteristics do not allow these compounds to be readily extracted by any organic solvent. Triethyllead and trimethyllead can be transferred as the undissociated chloride to an organic solvent, after saturation of the aqueous phase with NaCl [82, 84, 91]. To exclude possible interferences specific purification steps should be incorporated in the procedure [92], particularly in case of a final measurement which is not species-specific, as with graphite furnace AAS. For the determination of dialkyllead, a spectrophotometric method using 4-(2-pyridylazo)-resorcinol (PAR) has been adopted by a number of workers [82, 84, 93, 94]. The principle has recently been included in a chemical reaction detector coupled with HPLC [95, 96], thus allowing complete speciation, whereas the PAR method on itself can not discriminate between various dialkyllead products. Other complexing agents like dithizone result in a less selective extraction, as they also react with trialkyllead [84, 97, 98].

A real breakthrough in species-specific water analysis has been provided by the optimization of a sample treatment and enrichment procedure which allows a subsequent complete speciation by gas chromatography-atomic absorption spectrometry or related instrumental methods. Since the pioneering work on the extraction of ionic alkyllead species $R_nPb^{(4-n)+}$ as diethyldithiocarbamate complexes [99, 100] followed by derivatization to tetraalkylated homologues of the form $R_nPbR'_{4-n}$ using Grignard reagents (R'MgCl) [100, 101], the GC-AAS methodology has been widely accepted as the most versatile technique known today to address the complete speciation (TAL and ionic alkyllead) in water samples. Whereas the original method of Chau et al. [100] relied on butylation to render the compounds sufficiently volatile for gas chromatographic separation, others have similarly applied phenylation [102] or propylation [103], all appropriately assuming that only methyl- and ethyllead species can be expected in the environment. Further modifications have focused on the extraction and enrichment step, leading to a procedure which excludes the inconvenient presence of inorganic lead and which results in complete recoveries and superior detection limits, situated below the ng Pb l^{-1} level; for this purpose, however, the simultaneous determination of tetraalkyllead had to be sacrificed [85, 104]. Undoubtedly the flourishing of the GC-AAS methodology has greatly improved the quality and quantity of concentration data from the biosphere. A typical GC-AAS chromatogram of a rainwater sample is reproduced in Figure 4.

Recently, hydride generation procedures emerged which can be evaluated as a variant on the previous theme. Stimulated by the successful application of the prin-

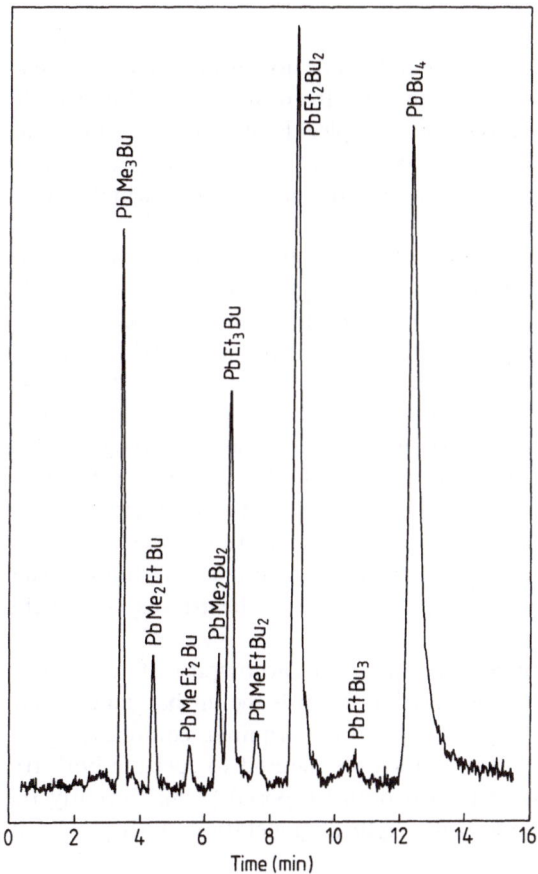

Fig. 4. Typical GC-AAS chromatogram of ionic alkyllead species in rainwater after extraction and butylation (Me = methyl, Et = ethyl, Bu = added butyl group). (From reference 199)

ciple for the speciation of arsenic and related elements [105], it was attempted to determine inorganic lead, tetraalkyllead, trialkyllead [106] and later also dialkyllead [107] using hydrogenation by sodium tetrahydroborate, cold trapping of the hydrides and subsequent introduction into a quartz furnace AAS. A procedure without trapping of the hydrides has also been described, involving many calculations by difference, and suffering from the inability to obtain individual speciation [108] Furthermore, speciation of lead and methyllead ions using in situ ethylation by sodium tetraethylborate has been realized [109]. Generally, however, the typical drawbacks of hydride generation procedures, like a modest reproducibility and abundant sources of interference, have hardly received attention; taking into account the relative instability of organolead hydrides, these aspects certainly have to pass a closer examination.

Finally, the development of electrochemical techniques for organolead speciation should be mentioned. Unfortunately, the inherent characteristics of differential pulse anodic stripping voltammetry allow only partial speciation into compound classes [110–112]. Moreover, the complex behaviour of organolead at mercury electrodes may give rise to inaccurate signals [113, 114]. Consequently, we feel that the reliability of electrochemical techniques for organolead analysis needs to be investigated with caution before their environmental application can be recommended.

Particulate Material (Air Particulate Matter, Dust, Sediment) and Biological Material

Speciation of organolead compounds in particulate and biological material consists basically of an extraction step to remove the analytes from the sample, followed by a quantification similar to that applied for water samples. From the extraction point of view, two approaches finally join the same way: solvent extraction directly into an organic solvent, and leaching of the species into an aqueous solution, which is then further subjected to solvent extraction.

The direct transfer into an organic solvent is the obvious method for the analysis of tetraalkyllead. Hexane is particularly suitable for the purpose: this has led to the characterization "hexane extractable lead", to indicate that fractions of compounds other than TAL may be collected in this way. Accordingly, the total lead determinations in the original procedures [86, 115–118] are now preferably abandoned in favour of species-specific methods [86, 119, 120]. As an alternative for solvent extraction, vacuum destillation at room temperature, followed by cryogenic trapping and extraction from the condensed liquid, has been proposed [121, 122]. As the recovery of the procedures has generally been monitored on spiked samples, which thus may contain the species in a form other than "natural", the question arises whether a comparable removal of TAL can always be assumed in environmental samples. This recovery problem, which is inherent to all complex matrices, holds true as well for the analysis of ionic alkyllead.

Extraction of trialkyllead into an organic solvent can be accomplished by "salting-out" procedures, using salts to shift the dissociation towards the uncharged covalent compound. Whereas originally tedious purifications were required, complicating the analysis [119, 123, 124], a more convenient and simple method to speciate both tri- and dialkyllead compounds as classes has also been proposed [125]. Recently the principle has been included in methods using hydride generation AAS [126] and HPLC-AAS [127] for the final determination.

The most promising strategy at present appears to be complexation of the species from an aqueous sample "solution". Since the first attempt in this direction using dithizone [128], most of the extraction procedures developed for water have been modified to allow analysis of the particulate fraction of ambient air, dust, sediments and biological material. Usually, recovering the analytes from solid matrices is not as straightforward as from water, requiring a more intensive treatment. The homogenized aqueous sample "solution" may be directly shaken with an organic solvent and a solution of the complexing agent, usually combined with salting-out agents [78, 129–133]. Some authors have preferred to expose the sample previously to a leaching procedure, the leachate then being treated in the same way as aqueous samples [85, 134–136].

Anthropogenic Sources of Organolead in the Environment

Since their introduction as antiknock agents in the combustion of gasoline, tetra-alkyllead compounds undoubtedly play a very important role in society. During the past decades only few organic chemicals have been produced in greater quantity [26]. Consequently the question has arisen about their possible threat to man and the human environment. Although detailed research projects focusing on the sources of organolead contamination came along with the debate, it remains difficult to ascertain quantitatively at present the specific routes through which the pollutants enter the biosphere. The variable composition of the antiknock blend mixture, the stepwise decreasing application during the past decade in some countries, and last but not least the evolution in engine and exhaust concepts in modern cars, are some of the factors which complicate straightforward characterization. The human impact on organolead pollution, however, is obvious and this chapter will evaluate these anthropogenic sources; further on we will deal with natural sources of organic lead, one of the major problems still waiting for further elucidation.

Tetraalkyllead as Antiknock Additive in Gasoline

Lead is added to gasoline in varying proportions of the five tetraalkyllead compounds in order to prevent the spontaneous premature combustion of the fuel mixture (knocking); this would give rise to a decrease in thermal efficiency and high mechanical strain on various parts of the engine. Although the mechanism of knock is relatively poorly understood, the technology of its control is well developed. The chief antiknock in general use in gasoline-type fuels has been tetraethyllead, applied from 1923 on. Since 1960, the other tetraalkyllead species, though intrinsically less effective, are usually blended together with TEL, as their increased volatility may be beneficial for the combustion of lighter hydrocarbon fractions [137]. The effectiveness of TAL as an antiknock agent relies on its ability to be readily oxidized to lead oxide which advances as a fine mist in front of the flame front in the combustion cylinder, scavenging the peroxy radicals which are considered responsible for the preignition reactions [27].

Because of the inclusion of scavengers (mainly 1,2-dichloroethane and 1,2-dibromoethane) to limit the deposition of lead in the exhaust, most of the exhausted lead is in the form of the halide salt PbBrCl or related double-salts; these then undergo further transformations in the atmosphere [138–140]. It has been estimated [139, 141] that between 50 and 70% of the input lead is emitted through the tail pipe, largely dependent on factors like the driving mode and the type of vehicle; the remainder is retained within the exhaust system and lubricating oil, and may partly be expelled by mechanical and thermal shocks.

The amount of additive emitted as organolead in the exhaust, i.e. the fraction which is not converted into inorganic lead, is relatively small. Estimates have been reported varying from 0.1% of the total lead in the exhaust [139] to about 13% [142], determined largely by the type of vehicle and the driving conditions. In addition to tetraalkyllead, appreciable amounts of ionic alkyllead species are apparently emitted, in both the gas and aerosol phases [142]. City driving will produce more unchanged TAL in the exhaust than highway driving; when starting or idling, the amounts emitted can indeed be appreciably higher than during driving at constant speed [142, 143]. On a winter day, Rohbock et al. [144] observed that R_4Pb made up 1.5 to 3% of the amount of particulate lead in the air during most of the day; during the evening rush hours from 4.00 to 6.00 p.m. this proportion increased to 11%. Evaporative losses from the fuel tank and carburettor, and crank case blow-by gases, which contributed significantly to emissions in older cars, can at present be assumed to be less important [5]. Taking this into account , Grandjean and Nielsen [145] put forward an average TAL emission of 0.7% of the input lead for a modern U.S. car of the seventies. A similar emission of incompletely burned fuel can be assumed for motorcycles, lawn mowers, outboard engines and others, but remains unconfirmed as yet.

Apart from direct engine emission, the commercial use of leaded gasoline involves other sources of TAL in the biosphere. Evaporative losses undoubtedly occur during fuel handling operations, at the moment of manufacture, transport and sale, as indicated by the elevated atmospheric concentrations of TAL sometimes found near gasoline filling stations [81]. Appreciably more tetramethyllead than tetraethyllead must be expected to evaporate because of its higher vapor pressure [26]; this is reflected by the ratio of both species in the surrounding air compared to the gasoline composition [81, 146]. Huntzicker et al. [141] have suggested that the loss of tetraalkyllead from filling stations constitutes about 1.3% of the total. Furthermore, accidental spillage or discharges may cause local contamination of water and soil, as historically witnessed by the Cavtat incident, in July 1974, when a shipload of about 325 tons of alkyllead antiknock compounds was dispersed in the Adriatic sea, of which fortunately a large part could be salvaged undamaged a few years later [35, 36]. Water and soil get also polluted through displacement processes acting upon atmospheric organolead, like dry and wet deposition, and eventually from the hydrosphere the compounds may settle into sediments. In this case no primary source is concerned (see later sections).

In summary, the total release of TAL in the biosphere due to its use as antiknock additive can apparently be estimated at about 2—3%, although there is no hard evidence to support this conclusion. Taking into account a worldwide consumption of refined lead for the manufacture of gasoline additives of about 240,000 ton in 1982 [5], this would mean that yearly some 5,000 ton organolead (expressed as Pb) enter the environment. As Western Europe is following the example set by the U.S.A.,

Japan and Australia by introducing legislation to further reduce vehicle pollution, requiring the availability and use of unleaded gasoline, this figure probably over-estimates the current emission.

As indicated in Figure 5, the decrease in the worldwide consumption of lead additives since 1974, the application date of the Clean Air Act in the U.S.A., can almost fully be ascribed to the imposition of more stringent conditions in that country. Consequently, a further drop should have occurred since 1982, the conversion to the unleaded gasoline market being a slow process dependent on the lifetime of old cars. Undoubtedly the curve will continue to decrease gradually: although the EEC compromise of March 1985 does not include a strict timescale, it requires the reduction of the maximum lead content of gasoline from $0.4 \, \mathrm{g \, l^{-1}}$ to $0.15 \, \mathrm{g \, l^{-1}}$ "as soon as appropriate" and the introduction of lead-free gasoline by 1989. The former measure has meanwhile been evaluated in the U.K. through sampling campaigns of particulate-bound lead in the air before and after the phase-down and has demonstrated a clear reduction, though somewhat smaller than theoretically expected [17, 147, 148].

The unpredictable, more dramatic part of the present evolution might be that manufacturers of antiknock fluids find an escape in selling the compounds to developing countries, where knowledge of lead toxicity and ecological impact has not yet resulted in governmental regulations. If that would happen the lead debate can be expected to start over once again.

When gasoline lead levels are lowered, or leaded gasoline is banned, the optimum octane number (which results in the lowest total energy consumption in car and refinery) is decreased. This fact creates a rush to other octane-boosting compounds: numerous products, e.g. methylcyclopentadienyl manganese tricarbonyl (MMT), ethanol, methanol, methyl tertiary butyl ether (MTBE) and tertiary butyl alcohol, have been tried with a variable degree of success [149]. Another point of view is the

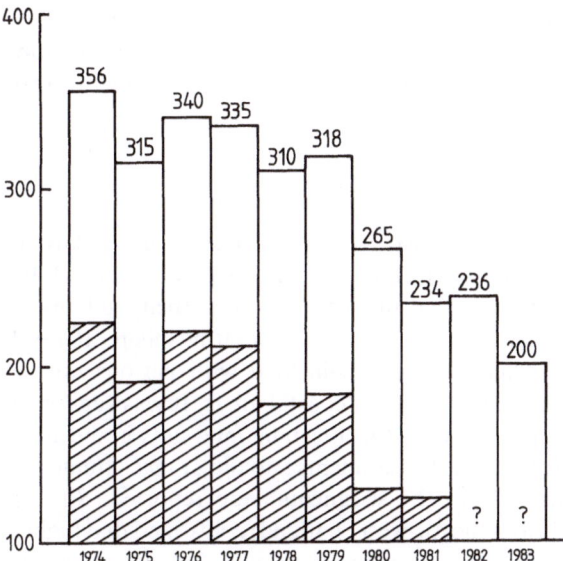

Fig. 5. Evolution of the annual consumption of lead in gasoline additives (10^3 ton Pb) in the Western World and in the United States (shaded part)

lean-burn technology development, which means more efficient engines with a higher air to gasoline ratio. Several companies, backed by the British government, feel that this may prove a more effective way of meeting the U.S. emission standards [18].

Other Commercial Uses of Organolead

Possible restrictions on the addition of lead to gasoline have long motivated manufacturers to search for new outlets for marketing the organoleads. The use for other commercial applications, however, remains as yet severely limited by environmental and public health considerations. In 1967 the market potential for these applications was still estimated to be about half the quantity of lead used for the production of antiknock agents. New markets which were suggested as being promising [26] consisted of toxicity-related applications (wood preservative, antifouling agent in marine paint, pesticide) as well as synthetic ones (polyurethane foam catalyst, polyvinyl chloride stabilizer). At present, however, only the introduction as alkylating agents in the manufacture of mercurial fungicides appears to have been realized to some extent [5], and consequently the effect of such applications on the environment is likely to be negligible.

Natural Production of Organolead in the Environment

A number of metals and metalloids may undergo methylation under environmental conditions through chemical or biological processes. Particularly the often more pronounced toxicity of methylated compared to inorganic metal species has stimulated much research work in this field. In a variety of studies it has been attempted to gather evidence to assess the environmental formation of tetraalkyllead. Whether in nature biomethylation of lead occurs, and to what extent, remains an unresolved problem as yet, largely due to the inconclusive character and non-reproducibility of many experimental data in this field. Furthermore, the possible natural production is likely to be a small-scale process relative to the current anthropogenic discharges of organolead into the biosphere, a fact which does not facilitate investigating the significance of biomethylation.

Environmental Evidence

Soon after Wong et al. [150] had first forwarded the idea of lead biomethylation, relying on the incubation of sediments, environmental evidence was claimed by Harrison and Laxen [151]. They referred to that process to explain their findings of elevated organic to inorganic lead concentration ratios in the atmosphere at rural locations in northwestern England when sampling coincided with periods during which the air masses had passed over "clean" sea and coastal areas. It was suggested that the source of the (unspecified) alkyllead were the intertidal mud flats in the area, and that tetramethyllead had been formed in the sediments; no direct measurements, however, of tetramethyllead evolution from the sediments were reported. Similar measurements made in the Outer Hebrides (north-west Scotland) confirmed that higher organic lead to total lead ratios may be found in maritime, unpolluted, air masses than predicted from urban source area measurements [152]. Recently, on the basis of a study on the atmospheric transformations of organolead, it was concluded that degradation of airborne TAL to the relatively stable trialkyllead during long-range transport could probably explain the enhanced ratios, making the hypothesis of a natural alkylation of lead unnecessary [80]. The unusually high ratios have also been ascribed to meteorological conditions, leading to a significant washout of particulate lead but keeping gaseous tetraalkyllead less affected [153].

The elevated concentration of organic lead found in marine fauna from presumably unpolluted water [83, 86, 117, 122, 125] has similarly led to suggestions on the existence of environmental or in vivo methylation of inorganic lead. The high lipid solubility of organolead, on the other hand, makes selective concentration of aquatic species a perhaps more plausible explanation [154]. Two Canadian groups have recently gathered additional indirect evidence to support the biomethylation idea, and more precisely a biomethylation starting from the ethyllead compounds as well as from inorganic lead. In an extensive monitoring program on the polluted river systems downstream from two alkyllead manufacturing plants in Canada, regularly significant amounts of mixed alkyllead species (methyltriethyllead and dimethyldiethyllead) and of trimethyllead were detected in living organisms and sediments, but not in water. As there was no apparent production or known usage of methyllead compounds in the area, it was believed that the ethyllead and inorganic lead species from the effluent would have been methylated.

An interesting new light on the question has also been shed by a series of measurements of tetraalkyllead in the air at rural continental sites in China [155]. No species could be detected, the detection limit of the procedure amounting to about 0.1 ngPb m^{-3}. As other vapor-phase alkyllead compounds were not monitored the possibility of a natural source of alkyllead cannot rigorously be excluded [156]. Nevertheless, the absence of tetraalkyllead in an environment with few anthropogenic inputs indicates that biomethylation of inorganic lead, if it occurs, might appear as a rather insignificant process in the atmosphere.

Simulation Experiments

During the past fifteen years, numerous attempts have been undertaken to address the biomethylation of lead on a laboratory scale, under more or less realistic conditions. Accordingly, the methylation in both environmental media and chemical systems has now widely been explored. Although extrapolation of the — often contradictory — results remains a controversial subject, the studies have certainly thrown some light upon the mechanism of the possible alkylation process.

It has been established that incubation of trimethyllead salts with marine sediments readily generates tetramethyllead, whether it concerns anaerobic [150, 157–162] or aerobic sediments [162]. Although originally a biological process has been proposed, based on the absence of alkyllead generation in sterilized media [150], gradually the foundations of this hypothesis became weakened. It was discovered that even sterilized sediments could yield tetramethyllead [160, 161], the addition of triethyllead chloride resulted in tetraethyllead and not the expected species methyltriethyllead indicating the absence of a transfer of methyl groups [157], the outcome of the experiments was only unambiguously positive when tetravalent lead compounds were al-

ready present. This evidence led many investigators to believe that a chemical disproportionation suffices to produce TAL, with sulfide ions enhancing the redistribution [157, 161, 162].

A curcial question, is whether divalent lead can also be methylated in nature. Wong
et al. [150] claim to have methylated lead nitrate and lead chloride, but not lead hydroxide, cyanide, oxide, or bromide, when incubated with certain sediments. Similar
observations were reported subsequently [158–160], whereas on the contrary other
extensive investigations with sediments ended up in a failure to repeat the experiments
[161–163]. Using bacterial cultures, Schmidt and Huber put forward a mixed process
in which both redistribution and biomethylation were proposed to be involved on
kinetic grounds [164, 165]. Reisinger et al. [162] critically evaluated the biomethylation
in sediments and bacterial cultures, relying on labeled and unlabeled substrates and
lead compounds to discriminate between both mechanisms; in no case a labeled
methyl group was incorporated into the tetraalkyllead compounds formed from trialkyllead, nor did Pb^{2+} evolve any methylated species. Recently, similar work of
Hewitt et al. [80] resulted in contradictory findings, showing the release of alkyllead
from sediments and the incorporation of labeled lead into the evolved tetramethyllead.

The alkylation in aqueous systems has also been thoroughly examined, but could
not unequivocally decide on the pathway either. The methyl-donor capacity of methylcobalamin and related compounds has been firmly demonstrated starting from di-
and trimethyllead [166–168]. Demethylation of methylcobalamin was also observed
with lead (IV) oxides, but nevertheless no organic lead could be detected [169], possibly due to the instability of monomethyllead. At several occasions reactions between
lead(II) salts and methylcobalamin in aqueous solution did apparently not produce
any methylated lead, or even demethylation of the substrate [157, 169–172]. Recently,
however, a successful methylation of lead(II) ions by a dimethylcobalt(III) complex,
which is a more powerful carbanion donor, was announced [168, 172]. The mechanism of a possible biomethylation may consequently involve a carbanion donor of
the methylcobalamin type.

On the other hand, experiments have also been set up based on a carbocation donor.
Ahmad et al. [173] have described a purely chemical alkylation of lead(II) salts using
methyl and ethyl iodide as effective alkylating agents. In agreement with a comment
which followed this paper [174], the reaction is now generally assumed to have been
catalyzed by the aluminium foil-wrapped stoppers of the vessels used, reducing the
lead(II) salts to finely divided metallic lead, which is known to be rapidly methylated
[175]. Interestingly, Jarvie et al. [175] have reported a reaction of lead(II) salts with
methyl iodide, but finishing at the trimethyllead stage. On the contrary, other attempts
with carbocation donors have failed to produce any alkyllead [168, 172].

In a pure aqueous solution with or without sulfur-containing compounds present
(Na_2S, cysteine) methylation of trimethyllead has repeatedly been observed [157, 158,
161, 162, 164]. It must probably be originating from a redistribution process, as inorganic lead does not yield detectable amounts of tetramethyllead in such media [157].

In summary, it is still unclear whether lead is methylated in the environment, to
what extent, and by what reaction mechanism. The biological methylation has been
observed by several groups, but, rather surprisingly considering the independent
nature of these observations, they are not consistently reproducible; a number of un-

successful searches have also been carried through. As the ecological consequences of chemically and/or biologically produced tetramethyllead would be the same, there is, from an environmental point of view, at present an urgent need to clarify the extent of lead conversion in the biosphere, rather than to attempt elucidating the exact mechanism. Undoubtedly analysts will continue to get attracted by the natural formation of alkyllead, one of the major unraveled mysteries which remain in the biogeochemical "cycle" of lead.

Concentrations of Organolead in the Atmosphere

From an analytical point of view, the numerous approaches to improve insight in the type of organolead species and their concentration levels in the atmosphere can be divided into two major classes. Many of the earlier attempts belong to what is commonly designed "total nonfilterable organolead" methods: they finally yield a single lead concentration, making no distinction between the various tetraalkyllead species, and include possibly sampled vapor-phase ionic alkyllead. Unfortunately, in the developing period of these techniques, often no reliable precautions were built in to avoid interference of inorganic lead escaping the filter. This led to concentrations which tend to be abnormally high when compared to more recent estimates [40, 176 to 178]. These uncertain values have been reviewed elsewhere [47, 49] and will be omitted here. The majority of atmospheric measurements relying on those non-speciating procedures, however, are likely to present a true total organolead figure.

The second and perhaps most interesting group of results can be traced back to the speciation methodologies which have been optimized during the past decade. They offer a detailed picture of the occurrence of tetraalkyllead species in the gaseous fraction of the air at urban and rural sites as well as at locations where a specific pattern might be expected (gasoline stations, parking garages, highways, tunnels). Information on the presence of TAL in the particulate matter fraction is scarce and will be discussed separately.

Whereas the significance of tetraalkyllead as an atmospheric pollutant is fairly well documented, the contribution of ionic alkyllead compounds has long been regarded negligible. Recent investigations using speciation techniques have raised doubts about this assumption, and currently efforts are focused on assessing the abundance of ionic alkyllead in the gaseous and aerosol fraction of the air. The second part of this chapter will be dedicated to an up-to-date status of research in this field.

Tetraalkyllead Species in the Gaseous Fraction

As apparent from Tables 5 and 6 which summarize the main data corresponding to urban and rural areas, individual measurements at specific sites cover a wide range. A far greater variability in the concentrations measured over short averaging periods reflects the importance of diurnal fluctuations. Typical daily patterns show distinct minima at night and elevated concentrations during traffic peak periods [53, 63, 75, 144]. Factors affecting roadside concentrations include the traffic density and the driving mode; for instance, individual 20 min kerbside samples alongside a set of traffic lights during a period when cold vehicles (with choke-running) were leaving

Table 5. Typical tetraalkyllead and total organic lead concentrations in the gaseous fraction of urban atmospheres

Location and date of sampling	Site*	No	Species**	Organic lead (ng Pb m^{-3})		Organic/Particulate Lead (%)		Ref.
				Range	Mean	Range	Mean	
Stockholm 1969	U	130	TML, TEL	40–3400	310		10	143
London 1973	U	4	org Pb	40–110	78	0.9–3.4	1.7	64
Frankfurt 1977	U	60	org Pb	2–170	37		7.7	144
Stockholm 1977	U	70	TML, TEL	140–2100	470			179
Toronto 1978	U	1	TML, ... , TEL		14		2.2	60
Antwerp 1978	U	5	org Pb	76–262	137	12–21	17	52
London 1979	U	14	org Pb	16–190	80	3.4–15	7.0	51
Copenhagen 1979	U	2	TML, TEL	185–195	190			70
Stockholm 1980	U	6	TML	47–77	61			70
Antwerp 1980	U	9	TML, ... , TEL	49–109	83	4.6–13	7.9	81
Glasgow 1980	U	58	org Pb	2.7–144	39	0.8–25	5.3	53
Lancaster 1984	U	3	TML, ... , TEL	50–90	64	1.9–3.5	2.8	75, 78
	U	3	TML, ... , TEL	155–220	178	3.2–4.5	3.7	77, 80
Colchester 1986	U	24	TML, ... , TEL	0.7–<25				185

* U = urban
** TML = tetramethyllead, TEL = tetraethyllead, org Pb = total organic lead

Table 6. Typical tetraalkyllead and total organic lead concentrations in the gaseous fraction of semi-urban and rural atmospheres.

Location and date of sampling	Site*	No	Species**	Organic lead (ng Pb m^{-3})		Organic/Particulate Lead (%)		Ref.
				Range	Mean	Range	Mean	
Frankfurt 1977	SU	30	org Pb	1–90	24		7.8	144
Antwerp 1978	SU	5	org Pb	8–20	13	2.6–13	7.9	52
Copenhagen 1979	SU	41	TML, TEL	1–84	27		7	70
Antwerp 1980	SU	7	TML, ... , TEL	3–14	7	0.6–3.4	1.9	81
Glasgow 1980	SU	10	org Pb	1–54	16	0.9–6.7	3.0	53
Lancaster 1984	SU	3	TML, ... , TEL	2–7	4	1.0–3.2	1.9	75
Colchester 1986	SU	24	TML, ... , TEL	< 2–< 19				185
Frankfurt 1977	R	4	org Pb	1–7	3		3.5	144
Lancaster 1977	R	33	org Pb	0.5–230	19	1.5–49	11	151, 51
Antwerp 1978	R	4	org Pb	≦ 8		≦ 7.5		52
Zealand, DK, 1980	R	4	TML	0.5–2.5	1.5			70
Antwerp 1980	R	6	TML, ... , TEL	0.3–3.9	2	0.1–0.7	0.5	81
Glasgow 1980	R	8	org Pb	1.6–6.5	3.9	1.5–64	14	53
Lancaster 1984	R	5	TML, ... , TEL	0.2–1.3	0.6	0.2–2.6	1.3	78
Lancaster 1984	R	100	TML, ... , TEL	0.2–6.5	1.9		ca. 2	80
Outer Hebrides 1984	R	24	TML, ... , TEL	≦ 0.25	ca. 0.1		ca. 0.6	80
Bantry Bay 1986	R	8	TML, ... , TEL	< 0.2–< 1.7				185

* SU = semiurban, R = rural
** TML = tetramethyllead, TEL = tetraethyllead, org Pb = total organic lead

a factory, ranged up to 2500 ng Pb m^{-3}, well above the typical kerbside values of about 100 ng Pb m^{-3}, when the traffic is free-flowing [50]. Concentrations further depend on the proximity to the vehicular emissions. Measurements in a central London street [51] have showed a 30–35% decrease in the TAL concentrations found, when the sampling height was increased from 5 m to 14 m above the street, with a similar decline for particulate lead. This correlation has been outlined in other studies as well [52, 53, 59]. Futhermore, meteorological conditions should have a profound influence on the levels measured, as should equally the gasoline composition in the area of investigation. In Antwerp, for example, a strong correlation of the atmospheric TAL levels with the local gasoline species distribution has been noticed [81]. Tetramethyllead is generally found to be relatively enriched in the atmosphere, presumably due to its higher vapor pressure and chemical stability. The five tetraalkyllead species can be detected in urban air in Western Europe [81, 75, 78], with tetramethyllead and tetraethyllead dominating.

The data suggest an average alkyllead concentration in urban air of some 50 to 100 ng Pb m^{-3}, typically representing 3–8% of the particulate lead level. Ambient air at rural sites shows a mean concentration of about 2 ng Pb m^{-3}. This appears to be 1–3% of the amount of particulate lead, taking into account only specific TAL methods. As the ratios of nonfilterable organic lead (or TAL for specific methods) to (filterable) particulate lead are less dependent on sampling site characteristics [47] and less ambiguous than ratios using total lead [62], they are generally preferred on the absolute levels to compare measurements. The lower ratio of TAL to particulate lead in rural areas, as deduced from Tables 5 and 6, can be explained by considering that during long-range transport of the anthropogenic tetraalkyllead pollutants part will decompose to trialkyllead or eventually inorganic lead. This obviously leads to lower ratios of TAL to particulate lead. Moreover, the elevated ratios occasionally obtained in some rural areas using total organolead procedures [151, 53] may indicate that a substantial part of the analytes measured there consisted of relatively stable ionic degradation products of TAL. Recent speciation work on these compounds apparently confirms this hypothesis. As alkyllead is probably scavenged less efficiently from the atmosphere through deposition processes than inorganic lead, the ratio of total organolead to particulate lead may rise and even exceed the ratio expected from urban source measurements. At suburban sites, the influence of distance-related decomposition processes should be less spectacular; Table 6 demonstrates indeed that similar ratios as in urban air are recorded, whereas the absolute levels average at about 20 ng Pb m^{-3}, intermediate between urban and rural concentrations.

The question arises how long these estimates, based on measurements performed till 1986, can still be considered valid, keeping in mind the progressive reduction of the maximum permissible lead content of gasoline in many countries. The particulate lead level in air follows a reduction in gasoline lead content quite closely [17, 144, 147, 148], but the effect on airborne organolead has, to our knowledge, never explicitly been investigated. Whether a sample (series) can be considered as representative or not remains another problem. It has been argued, for example, that photochemical smog in cities like Los Angeles and London, characterized by elevated ozone levels, may give rise to uncommonly low ratios due to R$_4$Pb decomposition [153].

By carrying out measurements at specifically chosen locations much has been revealed or confirmed about the sources of organolead in the atmosphere. In parking garages, concentrations amount to 1000–3000 ng Pb m^{-3}, being typically 15 to 25 % of that of particulate lead [40, 54, 143, 144, 179]. A much lower concentration (about 200 ng Pb m^{-3}) was found in the air in a car repair shop [81], suggesting that a dominant part of the R$_4$Pb in the air in parking garages originates from emission at starts and during idling. The concentrations of R$_4$Pb in the air near gasoline stations appear to be low compared to parking garage air levels, and yield organolead to particulate lead ratios similar to those in city areas [64, 70, 81, 179]. Notwithstanding this, occasionally ratios up to 30 % have been obtained [180]. During the filling of large fuel reservoirs at filling stations near Antwerp a surprisingly high concentration of TAL of 2000 ng Pb m^{-3} was detected 2 m away, with the same species distribution as the gasoline sold in the stations [81]. The importance of filling stations as point sources, consequently, is still not fully understood. At highways, levels are typically in the range 1 to 70 ng Pb m^{-3}, corresponding with 0.5 to 3 % of the amount of particulate lead [58, 144, 151, 181]. This low ratio, which is also found in tunnels, agrees with the observation that constant speed driving gives only limited TAL emission through the exhaust, due to a better thermal destruction.

Tetraalkyllead Species in the Particulate Matter Fraction

While efforts have been concentrated on the evaluation of tetraalkyllead species as gaseous air pollutants, the possibility that in ambient air the compounds may be partly adsorbed onto the aerosol fraction has up to now largely been neglected. Consequently information on this topic is scarce and fractionary. Edwards et al. [115, 182] first predicted the existence of particulate organic lead, based on laboratory experiments with simulated atmospheric dust particles. Two forms of sorbed lead could be distinguished: one form which was readily removed upon heating or by simple extraction with n-hexane, and a second form remaining, the former probably being tetraalkyllead and the latter one or more breakdown products.

Early estimates employing a non-specific technique (extraction-AAS) showed organolead concentrations between 6 and 29 ng Pb m^{-3}, or up to 1.2 % of the total particulate lead fraction [183]. Using a species-specific analysis, it became possible to point out the contribution of TAL species in aerosols. Reported values are generally extremely low and only close to traffic, occasionally the various TAL species are detected. De Jonghe et al. [81] reported aerosol TAL levels "well below 1 %" of the gaseous R$_4$Pb, without presenting more details. In four urban aerosol samples taken in Lancaster, U.K., the combined TAL content amounted to 28–760 pg Pb m^{-3}, or 0.002–0.05 % of the inorganic lead concentration [78, 184]. At a rural site in the vicinity, only one out of five samples yielded a detectable signal, being a tetraethyllead peak representing about 9 pg Pb m^{-3} or 0.03 % of the inorganic lead concentration [78]. A recent measurement campaign [185] in and near Colchester, U.K., confirmed that the fraction of tetraalkyllead adsorbed onto particulate matter is negligibly small; in only 2 or 3 of the 48 samples which were analyzed traces of a tetraalkyllead species occurred. Considering the detection limit situated below 0.4 pg Pb m^{-3} [185], this convincingly indicates that possibly adsorbed TAL is rapidly broken down or revolatilized.

As the sampling apparatus for determination of gaseous TAL in the atmosphere usually includes a prefilter to remove the aerosol fraction, it has to some extent been verified whether such a filter may remove TAL from the air stream. Under the actual conditions adopted, the filters did, apparently, not retain any significant amount of gaseous TAL [68, 186].

Ionic Alkyllead Species in the Gaseous Fraction

It has since long been postulated that the environmental decomposition of tetraalkyllead to inorganic lead involves tri- and dialkyllead salts as intermediates, and that accordingly their presence in the atmosphere is likely. Nielsen et al. [70] first realized that a dominant proportion of these compounds, and particularly trimethyllead, might occur in the gaseous state, based on the detection of high-boiling aromatic hydrocarbons as gaseous atmospheric pollutants [187]. Independently, the presence of ionic alkyllead in the gaseous fraction was at that time inferred from indirect evidence obtained in comparative studies based on simultaneous sampling with a species-specific TAL method and a non-specific "total" organolead method. In this way, De Jonghe et al. noticed that the former method invariably indicated lower concentrations than the latter (ca. 90%) in a comparison of four pairs of samples taken near a gasoline station [71, 81]. Similar studies afterwards confirmed this discrepancy, using a more extensive data set obtained in both rural and city areas [75, 78]. As it was felt that analytical inaccuracies could not account for the difference, an organic lead component in the gaseous fraction of the atmosphere other than TAL was suggested.

Attempts to identify positively this "excess" alkyllead fraction have remained unsuccessful until recently when Hewitt et al. [77, 79] proposed a suitable collection and quantification technique, based on trapping of the species in deionized water and subsequent extraction, as carried out routinely for water analysis. Interference from decomposing tetraalkyllead could not be excluded, although it was claimed to be slight in simulation experiments [79]. Up to now, the method which originally focused only on trialkyllead compounds has been applied to urban and, mainly, rural samples [77, 80, 185]; a compilation of the reported data is presented in Table 7.

It is tempting to conclude that gaseous ionic alkyllead may comprise an appreciable proportion of total gas phase alkyllead. The average contribution of gaseous trialkyllead at rural sites is striking, and may apparently amount to 20% or more. At a background urban site as well, individual concentrations of trimethyllead up to 20 ng Pb m^{-3} have been detected, leading to ionic alkyllead to total alkyllead ratios up to 55% [79]. Nevertheless, it looks like the entire difference between total alkyllead, measured by a non-specific method, and tetraalkyllead cannot be completely accounted for by the presence of gaseous trialkyllead; whether dialkyllead explains the remaining excess is still unclear. This shortfall is probable due to the non-quantitative collection and recovery of the triAL and diAL compounds; the collection efficiencies of Hewitt's procedure are indeed rather poor, ranging between 78% for trimethyllead and only 46% for diethyllead [79].

At present, it would be premature to draw any conclusions on the relative abundance of the various species. Further research on interference-free procedures and their application to a complete speciation of the various lead fractions in the atmosphere will certainly improve our knowledge on the significance of gaseous ionic alkyllead compounds as pollutants, and the transformations of organolead in the atmosphere.

Table 7. Typical ionic alkyllead concentrations in the gaseous fraction of the atmosphere

Location and date of sampling	Site*	No	TAL (ng Pb m^{-3})		TriAL (ng Pb m^{-3})	
			Range	Mean	Range	Mean
Lancaster 1984	U	3	155—220	178	15—21	18.3
Colchester 1986	U	6	2.9–16.1	8.3	<0.3–19.8	3.8
	SU	24	<1.9–<19.2		<0.7–<5.1	
Lancaster 1984	R	50	0.9–6.5	3.0	<0.5–1.9	0.6
Bantry Bay 1986	R	8	<0.2–<1.7		<0.6	

* U = urban, SU = semiurban, R = rural

Location and date of sampling	DiAL (ng Pb m^{-3})		Ratio of gaseous ionic AL to gaseous total AL (%)		Ref.
	Range	Mean	Range	Mean	
Lancaster 1984			6–12	10	77, 80
Colchester 1986	<0.5–5.1	1.5	0–55	33	79
	<0.8–<1.6				185
Lancaster 1984				ca. 20	80
Bantry Bay 1986	<0.7–1.1				185

Ionic Alkyllead Species in the Particulate Matter Fraction

Considering the relative insignificance of the tetraalkyllead contribution to the estimated total organolead fraction in an urban aerosol, the major part of the particle-associated organolead in the atmosphere could consist of ionic alkyllead species. The low concentrations, however, make a sensitive analytical methodology a prerequisite to verify the individual abundance of tri- and dialkyllead species in the particulate matter fraction. Recently this difficulty has been overcome, allowing the detection of ionic alkyllead in a wide variety of matrices, including aerosols. It follows that the original procedure of Harrison et al. [183] may have seriously overestimated the importance of the particulate organolead fraction. Instead of the 6–29 ng Pb m^{-3} reported, levels are probably situated one or two orders of magnitude lower.

The average aerosol in a residential area near Antwerp, Belgium, was found to contain about 200 pg Pb m^{-3} of ionic alkyllead compounds [135]. All major species, including the mixed methyl-ethyl products, were detected, the dominant species being diethyllead and to a lesser extent triethyllead. This pattern was correlated with the earlier measured [81] absence of tetraethyllead in the gaseous fraction of the air in the same area, this despite the prominent occurrence of that species in the gasoline used at that time; the minor stability of the tetraethyllead species was suggested to account for both observations. Remarkably, the dominant abundance on particulate matter of diethyllead, being the least volatile of the common species, was not confirmed in speciation work by Harrison et al. [78], probably due to the inability of their technique to recover dialkyllead species completely [78, 188]. Recently, however, they managed to optimize their sample treatment procedure [103], and the distribution of the species in aerosol samples taken subsequently at urban and rural sites near Colchester, U.K. [185] broadly agrees with the pattern found in Antwerp, though generally lower concentrations are present, below 30 pg Pb m^{-3}.

Summary

Although at present it remains speculative to outline clearly under which physical and chemical form organolead occurs in the atmosphere, some trends can be noticed from the previous survey. It appears that particulate matter plays a minor role in the atmospheric occurrence of organic lead. The typical concentrations of both tetraalkyllead and ionic alkyllead in the aerosol fraction (in the order of 5–500 pg Pb m^{-3}) look about three orders of magnitude smaller than the levels of the gaseous products. This could be accounted for by the volatility of organolead compounds, as indicated by a more frequent observation of the less volatile species in the particulate matter fraction; other factors like a faster degradation to inorganic lead in this matrix could however as well contribute.

Among the gaseous species, the ionic degradation products of tetraalkyllead may be more important than historically thought. Preliminary results illustrate that they might well represent 10–50% of the total alkyllead in the atmosphere. They probably result from atmospheric decomposition of TAL, whereas a controversial study [189] mentioning that significant amounts might also enter the atmosphere through direct exhaust emission has recently been confirmed [142].

With regard to the total levels of lead in the air [190] it seems obvious that the contribution of organic forms of lead is nonnegligible, possibly representing some 5–10%. A recent attempt to arrive at a complete speciation of atmospheric alkyllead points to the same conclusion [191]. It seems worth, therefore, to continue monitoring the concentrations and elucidating the pathways of organolead in the atmosphere, in view of the pronounced toxicity of these compounds.

Concentrations of Organolead in the Hydrosphere

From a physico-chemical point of view, the widely diverging properties of tetraalkyllead and ionic alkyllead compounds will manifest in a fundamentally different behaviour in the aqueous environment. Tetraalkylleads are volatile and water-insoluble and their presence in water is only transient. They may finally be partitioned into the lipids of living organisms, adsorbed onto particulates, volatilized to the atmosphere, or broken down immediately to ionic species. The ionic tri- and dialkyllead species are water-soluble. It might be anticipated, therefore, that wash-out processes transfer atmospheric organolead into the hydrosphere. Degradation of directly spilled liquid tetraalkyllead can also be expected to contaminate water with ionic alkylleads. Particularly the risk of bioaccumulation has stimulated much research in this field, which will be evaluated below.

Tetraalkyllead Species

Due to their nature, TAL species have but occasionally been detected in a variety of aqueous samples monitored for organolead. Their relatively high vapor pressures, nonpolar character and lipophilicity can indeed be expected to prevent them from occurring at elevated concentrations in the hydrosphere. When the impact of spilled TAL was studied at Otranto, near the wreckage of the Cavtat, a limited series of water samples was taken close to the deck load of alkyllead-containing drums during the recovery operations three years after the accident. Concentrations of chloroform-extractable lead amounted to 5–800 $\mu g \, l^{-1}$, and were assumed a reasonable measure for the alkyllead content [192]. Sea water collected at distances from 100 m to some kilometers from the wreck always gave total lead values lower than 0.1 $\mu g \, l^{-1}$, indicating that only localized pollution occurred. In another study the total lead values for surrounding seawater ranged from 1 to 15 $\mu g \, l^{-1}$ [193], revealing the lack of a reliable, sensitive and species-specific analytical technique at that time.

Since then, numerous specific TAL determinations have been carried out in a broad range of surface waters, including lake, river, estuarine and sea water. With detection limits situated at the 1–10 ng Pb l^{-1} level, none of them yielded detectable amounts of tetraalkyllead [186, 194, 154, 195] except for few negligibly small concentrations in Ontario lake water [86, 196]. In the essentially hydrophobic surface microlayer of the St. Clair river in Canada, tetraethyllead was found to be present, but not in subsurface water [197]. In bulk atmospheric deposition, occasionally TAL species have been recorded. In 12 out of 37 samples, which were left in the field for varying periods to allow a sufficient volume to be collected, TAL concentrations ranging from 1 to 92 ng Pb l^{-1} have been demonstrated [184, 198, 205]. In addition, highway drainage water was found to contain tetraalkyllead levels up to 40 ng Pb l^{-1}; in about one third of the 37 samples the species could be detected. It is interesting to note that Harrison et al. extracted the water directly in the sampling bottles without prior filtration. As adsorption of TAL on vessels and particulate matter may occur, and is largely unpredictable, rigorous precautions are indeed to be recommended if those species are being monitored in water [88, 188, 194]. Whether adsorption processes may explain why others failed to detect any TAL species in bulk deposition or road runoff [131, 199] is not certain. Conclusive evidence indicates that the hydrophobic tetraalkyllead compounds, if they are present in environmental water, will undergo a rapid and quantitative degradation to the more stable trialkyllead species [198, 200]; accordingly the sampling time may be an important parameter.

Ionic Alkyllead Species

Despite their hydrophilic character, about five years ago virtually nothing was known about the occurrence of the ionic degradation products of tetraalkyllead in the aqueous environment. Since then, a great deal of data has been produced, coming along with the breakthrough of speciation techniques enabling a sensitive determination. A survey of representative measurements is offered in Tables 8 and 9. They indicate that concentrations of ionic alkyllead in water rarely reach the μg Pb l^{-1} level.

From Table 8, there appears to exist a fair consistency between the absolute concentrations in bulk deposition, as determined at various locations in Western Europe.

Table 8. Typical ionic alkyllead concentrations in bulk atmospheric deposition.

Location and date of sampling	Site*	No	Trialkyllead (ng Pb l⁻¹)		Dialkyllead (ng Pb l⁻¹)		Ionic Alkyllead/Pb⁺⁺ (%)		Ref.
			Range	Mean	Range	Mean	Range	Mean**	
Antwerp 1983	U	4	58–330	152			0.02–0.09	0.05 (1)	201
	SU	2	45–88	67			0.03–0.07	0.05 (1)	
	R	1		28				0.10 (1)	
Antwerp 1984	U	12	58–153	93	16–225	86	0.05–0.14	0.08 (1)	202, 199
	SU	14	18–76	47	11–47	24	0.04–0.07	0.06 (1)	
Lancaster 1984	SU	7	17–362	130	<71		0.01–3.04	0.72 (2)	198, 205
	R	4	<14–72	25	<71		0.01–0.65	0.22 (2)	
Black Forest 1984	R	51	<2000–62000		up to >50			(2)	203
Antwerp 1985, snow	U	7	72–205	132	29–112	46	0.05–0.08	0.07 (1)	199, 204
							0.10–0.30	0.17 (2)	
	SU	3	25–61	47	4–26	18	0.04–0.05	0.04 (1)	
							0.10–0.15	0.12 (2)	
Dortmund 1985–86	U	9	18–116	56			6.0 –13.4	9.6 (1)	96
	SU	5	62–165	100					
Colchester 1985–86	U	5	41–124	79	14–165	62	0.40–0.49	0.43 (2)	185, 205
	SU	14	7–136	59	3–800	99	0.03–16.7	2.03 (2)	
Colchester 1986, snow	U	1		167		119		0.25 (2)	185, 205
	SU	1		30		20		0.23 (2)	
Antwerp 1986–87	SU	56	2–172	30	1–79	14	0.02–0.49	0.16 (2)	204

* U = urban, SU = semiurban, R = rural
** (1) based on total Pb⁺⁺, (2) based on dissolved Pb⁺⁺

Table 9. Typical ionic alkyllead concentrations in environmental water

Type of sample	Location and date of sampling	Site*	No	Trialkyllead (ng Pb l^{-1})		Dialkyllead (ng Pb l^{-1})		Ionic alkyllead/Pb^{++} (%)		Ref.
				Range	Mean	Range	Mean	Range	Mean**	
Road drainage	Birmingham 1977	U	49	< 100000		< 100000				82
Road drainage	Antwerp 1983	SU	3	70–140	113				1.08 (2)	201
Road drainage	Burton-in-Kendal UK 1984	R	37	< 7–148	38	< 70		0.01–4.93	0.58 (2)	198
Road surface	Antwerp 1985	SU	3	660–1470	990	48–83	68			134
Road surface	Antwerp 1985	G	1		4131		159			134
Road surface	Colchester 1986	U	1		685		1240		19 (2)	194
Road surface	Colchester 1986	SU	1		66		< 11		0.21 (2)	131
Road surface	Colchester 1986	R	1		15		6		0.06 (2)	131
Road surface	Canada 1986	U	2	280–400	340	360–470	415	< 0.01–0.03	0.02 (1)	201
Road surface	Canada 1986	G	1		160		1800			195
River/Lake	Antwerp 1983		16	< 20						154
River	Canada 1983		13	< 8–400	115	< 8–80	22	< 0.8–29.4	7.1 (1)	195
River			6	< 8–390		< 8–150				154
River, surface	Canada 1983		10	< 8–1760	349	< 8–150	50	< 0.9–25.5	6.0 (1)	195
microlayer			8	220–540		140–610				154
River/Lake	Antwerp 1984–85		8	4.4–9.4	6	< 0.7–1.3	< 1			199
River/Lake	Dortmund 1985–86		2	< 35						96
River/Lake/Sea	Colchester 1986		1	< 1.8		< 1.4				194
River	Colchester 1986		1		0.8					194
River/Lake	Antwerp 1986		19	2.5–12.5	7.2	< 0.2–2.8	0.7			136
Sea	North Sea 1986		6	0.2–2.1	0.8	< 0.2				136
Harbour	Dortmund 1986		1		163				2.4 (1)	96
Harbour	Colchester, 1986		1		147		219		3.7 (2)	194
Industrial effluent	U.S. 1982		1		24000				4	101
Canal (near alkyllead plant)	Liverpool 1982		7	2900–92300	39000	2800–7700	4200			215
Tap water	Dortmund 1986		1	< 35						96
Tap water	Great Britain 1986		11	0.4–3.6	1.4	< 0.3–1.6	0.2	< 0.3		194
Tap water	Belgium 1987		7	0.3–7.1	2.7	< 0.7–2.2				206
Ground water	Colchester 1986		2	< 1.8						194
Ground water	Antwerp 1987		1		2.0	1.4	0.7			206

* U = urban, SU = semiurban, R = rural, G = near gasoline station

** (1) based on total Pb^{++}, (2) based on dissolved Pb^{++}

The only major exception is a series of trialkyllead measurements by Faulstich et al. [203] who reported that 14 out of 51 rainwater samples from the Black Forest, FRG, contained trialkyllead concentrations ranging from 2 to 62 μg Pb l^{-1}. As the highest amount detected was only three times lower than the concentration of Et$_3$PbCl that is toxic for neuroblastoma or soybean cells, and accumulation by needles of coniferous trees had been demonstrated [207], it was concluded that the continual exposure of trees to rainwater containing trialkyllead salts might contribute to the European forest damage.

Their point of view soon became heavily criticized, both on analytical and environmental grounds [208–210]. It has been claimed that the biochemical assay employed, which is based on the in vitro inhibition of tubulin polymerization in pork brain, is unspecific for ionic alkyllead, and subject to interferences by other constituents of rainwater. With regard to this, Faulstich et al. later stated [211] that they cross-checked their analytical method with a reliable chemical procedure. Independent analysis using a speciation methodology broadly confirmed the results but pointed to a possible contamination problem [212]. In an attempt to explain the discrepancy with other reported concentrations Faulstich et al. stressed that only occasionally toxic events were recorded, and that they were confined to small areas. In this context, it is worth noticing that an independent report equally points to atmospheric lead as a contributing factor in the forest decline syndrome, based on a correlation between symptoms and total lead content in two spruce stands [213]. Apart from analytical considerations, other factors have raised doubts about the hypothesis of the German research team, like the absence of damage symptoms at urban sites where the ionic alkyllead levels are expected to be highest, and the emergence of the phenomenon at a moment when the gasoline lead level has been drastically reduced compared to a decade before. In summary, the proposition that ionic alkyllead occurs at toxic concentration levels in rain is, in our view, at present insufficiently documented to consider it proven. None of the ca. 120 rain water samples collected so far at other locations and by other groups has yielded a total ionic alkyllead concentration exceeding 1 μg Pb l^{-1}, and this difference of about two orders of magnitude as compared to the Black Forest measurements makes the latter investigation at least dubious.

Although the other studies more or less agree on the absolute concentrations of ionic organolead, suggesting an average level of 50–200 ng Pb l^{-1} in rain, the reported values for the ratio of ionic alkyllead to inorganic lead are far more scattered. An evaluation of this factor is difficult, as in part of the investigations total inorganic lead has been monitored, whereas in other cases the dissolved inorganic lead (i.e. the fraction which is not retained by a 0.45 μm filter) has been preferred. The ratio of ionic alkyllead to total inorganic lead has been found to be remarkably constant near Antwerp, and amounts to 0.05–0.10%. Ratios based on dissolved inorganic lead are generally smaller than 1%, though occasionally ratios up to 17% have been measured. Particularly in some samples taken near Dortmund, constantly high ratios have been obtained [96]; the abnormally low inorganic lead concentration in this rain water (300–900 ng l^{-1}), if not erroneous, should explain this deviation.

Taking into account the variety of parameters which may influence the organolead composition of rainwater (gasoline composition, photochemical activity in the atmosphere, intensity and duration of rainfall, length of sampling period) it seems not surprising that correlations can hardly be outlined. No consistent seasonal variation

has been observed in two long-term sampling campaigns [204, 205]. From the Antwerp data, it is tempting to conclude that the highest concentrations in rainwater
coincide with the highest atmospheric TAL levels. This site-dependency illustrates
the importance of deposition processes as sinks for atmospheric organolead. The wet
deposition rate for alkyllead at a suburban site has been estimated at 56 μg Pb m^{-2}
yr^{-1} [205] and 32 μg Pb m^{-2} yr^{-1} [204] respectively, using funnel-in-bottle samplers
which also may collect a small but unquantified dry deposition component. Alkyllead
is scavenged less efficiently from the atmosphere through wet deposition than inorganic
lead, as is tempting to conclude from the higher alkyllead to inorganic lead ratios in
the air than in rainwater.

Concerning the abundance of the individual species, it is obvious from Table 8
that the amounts of dialkyllead in rainwater are of the same magnitude as these of the
trialkylleads. In addition to trimethyllead, dimethyllead, triethyllead and diethyllead,
also mixed methyl-ethyllead species are regularly observed. They can be assumed to
be present in the rain-out, as the precursor mixed tetraalkyllead species have been
systematically measured in the gaseous fraction of the air.

Repeatedly peaks have been recorded which were ascribed to RPb^{3+} species, based
on the chromatographic retention characteristics [199, 185, 205]. These peaks have
meanwhile been detected in almost all environmental matrices. The existence of monoalkyllead in the environment, however, is highly improbable due to its known instability. The apparent existence of these species in the analysis may very likely be due
to an analytical artifact, as indicated independently by two recent evaluations of the
extraction-derivatization-GCAAS procedure [214, 131]. If so, they possibly arise
from rearrangement of other species. The process manifests itself on a limited scale
(well below 10 %) in the Antwerp measurements [214]. Harrison and coworkers, however, have reported surprisingly high relative amounts of the apparent monoalkyllead
species, representing up to 2.5 times the sum of the concentrations of the other ionic
alkyllead species [194, 185, 205]. As long as an unambiguous proof of the environmental existence of monoalkyllead or the mechanism of its formation is lacking, it
must be kept in mind that such predominant signals might seriously affect the reliability of the levels of the other species. In the tables, the apparent monoalkyllead
species have been included in the total ionic alkyllead content, and the most dubious
measurements have been omitted.

Another aqueous matrix which has received much attention during the past few
years is drainage or surface water from roads. Particularly in samples collected directly
from the surface, the total ionic alkyllead level may exceed 1 μg Pb l^{-1}, as shown in
Table 9. The occurrence of the compounds in road water tends to be correlated with the
emission sources of tetraalkyllead; at urban sites and especially near gasoline stations
the ionic degradation products are clearly more abundant than in suburban and rural
areas. In runoff water collected during extensive rain showers, the alkyllead composition of the average rain water seems to be reflected, rather than a contribution from
road wash-off [198].

As apparent from Table 9, river, lake, sea, ground and tap water generally contain
but minute quantities of ionic alkyllead (10 ng Pb l^{-1} or less). In locally polluted river
water, however, sampled downstream from an alkyllead plant in Canada, levels up
to 470 ng Pb l^{-1} were detected, and the compounds were found to be even enriched
in the hydrophobic surface microlayer, which contained up to 1910 ng Pb l^{-1} ionic

alkyllead [154, 195]. No methyllead was detected in any of the samples, in agreement with the exclusive manufacture of tetraethyllead in the plant. Other industrial effluent samples and few harbour measurements confirm that wastewater from alkyllead manufacturers and users may pollute the aqueous environment with significant amounts of organolead. In both unpolluted and polluted water samples, the preponderance of trialkyllead compared to dialkyllead species is striking. As this tendency does not hold true explicitly for rain and road water, it seems unlikely that mere stability considerations would be involved in a sound explanation. Finally, it is worth mentioning that tap water contains similar concentrations of alkyllead as river or ground water, although this is normally submitted to purification before it is declared potable. Thus, at least part of the species may escape removal from water through the usual water treatment procedures.

Concentrations of Organolead in Dust, Soil, Sediment, and Biological Material

Despite the sometimes impressive obstacles which have to be overcome to arrive at a successful speciation of alkyllead in solid matrices, historically analysts have not refrained from facing the determination of these compounds in, for example, sediments and tissues. The fact that sediments were considered as the ultimate sink for aqueous organolead and at the same time a suitable medium for biomethylation, made them a popular study object, whereas biological tissues enjoyed a continuous interest for they could reveal the mechanism and extent of the toxicity of organic lead and the significance of a possible bioaccumulation. In many investigations, unfortunately, applications were limited to few "illustrative" samples. This section will attempt to summarize the environmental data and discuss their implication for man and nature.

Alkyllead Species in Dust, Soil and Sediment

The possible role of particulate material in the cycle of lead in the biosphere has convincingly been recognized in laboratory studies based on air and water analysis techniques, where adsorption of organolead on dust as well as on suspended particles in aqueous systems has been demonstrated [87, 88, 115, 182, 215, 216]. Since a few years, speciation methods allow to gather direct information on the concentration of alkyllead in dust, soil and sediment samples. The most important data are compiled in Table 10; the aerosol fraction of the air, which is also a particulate matter matrix, has been discussed in a previous section.

The alkyllead levels found in road dust samples look fairly consistent, with exception of an old report which probably overestimates the association with dust, and point to an average concentration of 10–100 ng Pb g^{-1}. This is negligible compared to the inorganic lead content, which typically amounts to ca. 1000 µg g^{-1} [116, 190]. Near gasoline stations, higher pollution levels can be expected, a similar trend as observed in road water, and which is again confirmed in the road drainage sediments. Soils do apparently contain but traces of ionic alkyllead, with exception for one sample in which the combined alkyllead concentration exceeded 200 ng Pb g^{-1}. In both dust and soil samples, tetraalkyllead is only exceptionally detected.

Table 10. Typical organolead concentrations in dust, soil and sediment

Type of sample	Location and date of sampling	Site*	No	TAL (ng Pb g^{-1})		TriAL (ng Pb g^{-1})		DiAL (ng Pb g^{-1})		Ref.
				Range	Mean	Range	Mean	Range	Mean	
Road dust	Lancaster 1976	U, SU, G	9	600–7400	3500***					116
Road dust	Lancaster 1985	U	1	< 7.4		< 2.6		< 14.0		188
Road dust	Antwerp 1985–86	SU	3			3.2–23.8	10.7	9.7–109.2	45.3	134
Road dust	Antwerp 1985–86	G	1				39.7		307.4	134
Road dust	Colchester 1986	G	5	< 34–52	10.5	1.3–27.0	8.0	0.4–33.0	9.5	132
Road dust	Canada 1986	U	1	< 34			8		42	131
Road dust	Canada 1986	G	1				29		166	132
Roadside soil	Colchester 1986	U	2–5	< 34–146	36.5	< 0.2–47.0	23.5	< 0.2–43.0	21.5	132
Roadside soil	Canada 1986	U	2			0.7–1.2	1.0	4.0–10.0	7.0	131
Roadside soil	Antwerp 1987	SU	3			0.1–0.6	0.4	0.2–2.2	1.1	206
Road drainage sediment	Birmingham 1977	U, SU	10	50–520	273	< 2000				82
Road drainage sediment		G	4	5000–100000	46900	2000–6100	2525			82
Road drainage sediment	Burton-in-Kendal UK 1984	R		< 7.4		< 2.6		< 14.0		198
Lake sediment	Canada 1980		3	0.5–160	88**	0.03–1.08	0.53	< 0.04–0.22	0.10	86
Lake sediment	Antwerp 1986		11			< 16–275	56	< 16–22	1.5	136
River sediment	Canada 1983		44	< 24–1290	124					195
River sediment	Canada 1983		7	< 8–290		< 8–910		< 8–1560		154
River sediment	Antwerp 1986		7			0.14–0.50	0.32	< 0.04–0.14	0.06	136
Sea sediment	Essex 1986		3			< 0.1–0.3	0.13	< 0.1–1.1	0.40	132
Sea sediment	Ireland 1986		1	< 0.2–0.4	0.2		0.20			132
Sea sediment	North Sea 1986		3			0.04–0.07	0.05	< 0.04	< 0.14	136

* U = urban, SU = semiurban, R = rural, G = near gasoline station

** lead fraction extractable in hexane

*** lead fraction extractable in cold ammoniacal methanol

The low ratio of organic to inorganic lead in dusts and soils may indicate a limited stability of organolead in these matrices. Preliminary studies on the degradation of ionic alkyllead species in some samples, however, have not substantiated this assumption [206]. A more probable explanation might be that the dominant proportion of the alkyllead salts remains gaseous with a minor fraction condensing and remaining onto particulate matter. The least volatile species, like diethyllead, seem indeed the most abundant in dust and soil, similarly as in aerosols.

The organolead levels recorded in sediments, as summarized in Table 10, can be split up in two groups, a first consisting of values for locally polluted sediments and another group of concentrations belonging to more general sites. This division closely follows the one which has been outlined in the series of natural water measurements. Typical river, lake and sea sediments may show a variety of organolead species, but in concentrations which are barely detectable, their sum lying usually below 1 ng Pb g^{-1}. The data set is too limited to enable a clear insight in the significance of the possible interaction between natural water and its sediments. Interestingly, downstream from an alkyllead manufacturing plant in Canada, levels have been measured which regularly fell in the range 100–500 ng Pb g^{-1}, with occasionally concentrations exceeding 1 μg Pb g^{-1}. Between 1982 and 1986, they were found to have dropped gradually from about 1 μg Pb g^{-1} to, finally, nondetectable amounts after the closure of the plant in 1985, consistent with the improved reduction of alkylleads in the effluent.

Surprisingly, impressive concentrations of trimethyllead (< 8–275 ng Pb g^{-1}), dimethyldiethyllead (38–263 ng Pb g^{-1}) and methyltriethyllead (< 8–168 ng Pb g^{-1}) characterized the distribution of alkyllead in the sediments sampled in 1983, both upstream and downstream from the plant, and made up the total lead content of the sediment in several cases [195]. As only ethyllead compounds were assumed to be produced and applied in the area, it was claimed that biomethylation processes should have generated the observed species distribution. It is remarkable, that one year later sediments sampled near the same downstream location contained exclusively the ethyllead compounds which could be expected, considering the local anthropogenic sources; in addition, their inorganic lead content was appreciably higher than in 1983. In 1986, the concentration of inorganic lead again reached about the 1983 level and almost no alkyllead was recorded. During the entire monitoring program, fish appeared to have accumulated alkyllead, but the proportion of methyl-containing species was generally small; in 1983 they occurred in only 30 out of 166 samples. Consequently, any correlation which might help to explain the observed pattern in 1983 sediments seems to be lacking; attributing it to some biomethylation process is a possible explanation, but needs further elucidation.

Alkyllead Species in Biological Material

Attempts to determine the alkyllead contamination in biological material have up to now mainly focused on three types of samples: fish and other aquatic organisms, birds, and brains, urine or blood from (mainly exposed) humans. In Tables 11 and 12, the obtained results have been put together. Although the reliability of reported values may critically depend on the analytical method employed, a thorough discussion on this matter is complicated and beyond the illustrative scope of this review. Most of the procedures applied were claimed or proven to be species-specific, though the identity of the extracted compounds was not always assessed. Sample preparation is commonly carried out using a tissue solubilizer (tetramethylammonium hydroxide) or enzyme hydrolysis using mixed enzymes, lipase and proteases.

A quick inspection of the organolead concentrations in the wide variety of fish analyzéd in the course of the past decade (Table 11) leaves no doubt about accumulation of these pollutants in living organisms. In the polluted St. Lawrence river system downstream from a plant producing tetraethyllead, a carp containing 139 µg Pb g^{-1} of alkylleads has been caught, and levels of 1–10 µg Pb g^{-1} occurred regularly [195]. The organolead burden of local fish decreased significantly between 1981 and 1987, in close agreement with the trend observed in water and sediments. At the end of the period, after closure of the plant, only occasionally the typical "background" values were exceeded, which can be estimated in the low ng Pb g^{-1} range as suggested by a number of measurements in relatively unpolluted areas.

Compared to the whole fish, alkyllead levels used to be more elevated in intestines. The spectrum of tetra-, tri- and dialkyllead species corresponded most closely to the spectrum in water, but not to that in sediments. On the average, some 50–75% of the total lead in the fish samples consisted of alkylleads, with tetraethyllead and triethyllead as most common (and expected) forms. A similar picture and evolution of organolead levels has been observed in fish from another contaminated area, and in macrophytes. Clams were also studied, and proved to be suitable indicators when placed in cages in the river bed, allowing to pin-point the influence of the source [154].

Unexpected bird mortalities in an important wader and wildfowl wintering area on the Mersey estuary were the first sign that alkyllead may accumulate in birds. As casualties exhibited typical lead intoxication symptoms, the chemical form and concentration of this metal was examined. This revealed that dead birds contained sufficient lead to have caused the mortalities, and dominantly under the form of ionic alkyllead, mainly in kidney and liver. The presence of the pollutant was consistent with the manufacture and use of alkyllead on the river, and was also confirmed in the invertebrate preys of the birds. The concentration of ionic organolead in live waders in 1981 seems only few times larger than the amounts detected in pigeons from nearby Liverpool, as indicated in Table 12. Birds monitored in Canada, on the other hand, have apparently much lower alkyllead burdens, below 10 ng Pb g^{-1}. As for these measurements an entirely specific GC-AAS methodology was employed, the data may represent a reliable estimate of typical "background" contamination levels in bird tissue.

The examination of Herring Gulls, which are considered a valuable regional contamination indicator [130], has resulted in an interesting species distribution pattern. Trimethyllead was found an ubiquitous, and frequently the most abundant, species, despite the assumed absence of anthropogenic methyllead emissions. In addition, dimethyllead and mixed methyl-ethyl lead compounds were detected. These species also occurred in a type of saltmarch snail [220]. In this context it may be recalled that, in their extensive survey of fish and sediments in Canada, Chau et al. have regularly observed the mixed organolead species. In an early publication, they also mentioned significant concentrations of trimethyllead in several samples [129] which were however, apparently, omitted and excluded from the total alkyllead level in the final report [195]. These independent observations make an analytical artifact improbable, and

Table 11. Typical organolead concentrations in fish and aquatic organisms

Location and date of sampling	Sample	Part	No	Organic lead (ng Pb g⁻¹) Range	Mean	TAL (ng Pb g⁻¹) Range	Mean	Ionic alkyllead (ng Pb g⁻¹) Range	Mean	Ref.
Halifax, Canada 1976	Fish	Various	7	10–4790	745					117
Mediterranean, Italia 1977	Mussel		4	9690–47900	22350					122
North sea, Belgium 1978–79	Fish	Fillet	75	10–70						118
		Liver	12	240–480						
Ontario, Canada 1980	Fish	Whole	17	2–72	17	< 1–16	4.3			196
			90	< 2		< 0.5				
N.W. Coast, England 1981	Fish		3			< 20		< 20–50	27	125
St. Lawrence river 1981	Fish	Intestines	28			< 24–62900	3800	< 16–100500	7130	195
Canada, (downstream from alkyllead plant)		Whole but intestines	28			< 24–73100	6680	< 16–68700	4490	129 154
1982		Intestines	29			< 24–27100	2310	< 16–7700	1000	
		Whole	45			< 24–58600	1930	< 16–4300	560	
1983		Whole	26			< 24–6100	950	< 16–5200	850	
1984		Whole	109			< 24–3900	230	< 16–2300	190	
1986		Whole	25			< 24–10500	550	< 16–1700	130	
1987		Whole	16			< 24–3600	220	< 16–740	63	
St. Lawrence river, Canada (upstream from plant) 1984	Fish	Whole	41			< 24–930	35	< 16–1400	48	195
Ontario lakes, Canada 1983	Fish	Whole	16			< 24–32	2	< 16–63	4	195
St. Lawrence river, Canada 1982	Clams		2			< 24–53	27	< 16–280	140	195
1982	Macrophytes		3			< 24–21200	7100	< 16–670	270	195
1983	Macrophytes		11			< 24		< 16–190	18	
1984	Macrophytes		13			< 24–190	22	< 16–240	35	
St. Clair river, Canada, downstream from alkyllead plant 1984	Fish	Whole	22			< 24–1130	130	< 16–390	92	195
1987	Fish	Whole	11			< 24–118	11	< 16–118	19	195

Table 12. Typical organolead concentrations in animal and human tissue

Location and date of sampling	Sample	Part	No	TAL concentration (ng Pb g^{-1} or ng Pb ml^{-1}*) Range	Mean	Ionic alkyllead concentration (ng Pb g^{-1} or ng Pb ml^{-1}*) Range	Mean	Ref.
Liverpool, UK (Mersey estuary, near alkyllead plant)	1979 Dunlin (dead)	liver	10				10100	217
		kidney	10				13800	
		muscle	10				4050	
		brain	10				7690	
		bone	10				1040	
	1981 Dunlin (alive)	liver	10				720	
		kidney	10				1180	
		muscle	10				330	
Liverpool, UK	1979 Invertebrates					200–1000		218
	1981 Pigeons	kidney	7	< 100			310	219
		brain	7	< 90			20	
Bridport, UK (rural area)	1981 Pigeons	kidney	7	< 130			80	219
		brain	7	< 130			< 10	
Great Lakes, Canada 1983	Herring Gulls	liver	8			3.0–8.3	5.5	130
		kidney	8			5.0–17.9	9.2	
	Chicken	liver	4				ca. 3.1	
		kidney	4				ca. 6.4	
		egg	10			< 0.13–1.29	0.43	
St. Lawrence river Canada	1983 Ducks	liver	10			0.3–8.1	2.0	220
		kidney	4			2.0–10.0	5.8	
		muscle	2			0.8–4.3	2.5	
		brain	2			1.0–3.4	2.2	
	1985 Snails		12			2.2–39.5	20.5	
Lodz, Poland	1965 Intoxicated humans	liver	3			7500–22500 (1)		124
		kidney	3			4500–17500 (1)		
		brain	3			4000–14000 (1)		
		urine	5			46–230 (1)		
		blood	2			< 200–2400 (1)		

Location	Year	Subject	Sample	n					Ref
Copenhagen, DK	1978	Humans (residents)	brain	22			< 8–50		221
Japan	1974	Intoxicated humans	urine	8			13–183	53	222
Lodz, Poland	1983	Intoxicated humans	urine	7			65–236000 (2)		223
Umea, Sweden	1983–86	Humans (occupationally exposed)	blood	9	5–27	9.7			120
				5	< 2–3	1.2	< 3–5 (1)	1.0	133
		Humans (residents)	blood	5	< 3				120

(1) only triAL
(2) only diAL
*   for blood and urine samples

suggest a possible biomethylation of lead and ethyllead species as alternative for a so far overlooked anthropogenic pollution source in Canada.

As contrasted with animal tissues, there is a dearth of knowledge about the occurrence of organic lead in plant material. After the recent report of Faulstich et al. that points to a contribution of ionic alkyllead in the European forest decline [203], an attempt has been undertaken to speciate these pollutants in grass and tree needles [204]. In 15 samples collected at various sites, the total ionic alkyllead level amounted to 1–10 ng Pb g^{-1}. Five grass samples cut close to a gasoline station near a traffic light-secured road intersection during the winter 1985–1986 contained more elevated concentrations, between 17 and 100 ng Pb g^{-1} [204]. Consequently, it cannot be precluded that significant amounts of ionic alkyllead may be associated with plant material, either really taken up or superficially adsorbed.

Organolead speciation in humans has only been carried out after acute intoxication, and recently also in occupationally exposed workers, such as tank cleaners. Blood levels of organic lead compounds up to about 25 ng Pb ml^{-1} were characteristic for the latter group, representing up to 15% of the inorganic lead content [120]. In urine from intoxicated persons dialkyllead is the dominant form of organolead [222, 223]. Analysis of brain samples from residents of Copenhagen, and the surrounding suburbs and rural areas, has revealed the presence of (presumably) trialkyllead in human brain at a concentration level of < 8–50 ng Pb g^{-1} [221].

Sink Processes for Organolead in the Environment

To allow a comprehensive evaluation of the impact of organolead on the environment, knowledge on sources and occurrence of the various species needs to be completed by detailed investigations into the transformation and displacement processes which eventually remove them from the biosphere. Although the field has been explored to some extent, there remains a lack of environmentally significant data. Convincing evidence indicates that in every environmental matrix TAL compounds are eventually converted into inorganic lead through triAL salts; diAL salts are likely to be involved as intermediates. Monoalkyllead salts are apparently too unstable to permit their isolation, but have been supposed to be intermediates in a final conversion step to inorganic lead [26, 32, 165, 224]. Their occasional appearance in GC-AAS determinations, derivatized to the form of mixed tetraalkylated compounds, can almost certainly be ascribed to a redistribution caused by an analytical imperfection [214, 131]. On the stability of the species and the pathway followed in the transformation reactions and on the importance of the displacement processes between different compartments of the biosphere, information is until now fractionary. It will be attempted below to structure and update the knowledge on the sink processes from the atmosphere and the hydrosphere.

Atmospheric Sink Processes

The principal reaction pathways of tetraalkyllead breakdown under atmospheric conditions are homogeneous, and typical of those applying to hydrocarbons in general. Degradation is due primarily to attack by reactive species generated by photochemical activity like hydroxyl radicals and ozone; in addition the small absorbance of TAL in the tropospheric solar UV region implies a pathway for direct photolytic decomposition. The rate constants for the photolysis of tetramethyllead and tetraethyllead have been estimated to be 8 and 26% hr^{-1}, respectively, at a solar zenith angle z of about 40°, and 2 and 7% hr^{-1}, respectively, at a solar zenith angle z of about 75° [216]. This corresponds roughly to bright sunshine at noon in summer and in winter, respectively, at 55° N latitude. In pure air under conditions of darkness, decomposition rates were found to be much lower, 0.2% hr^{-1} and 0.7% hr^{-1} for the respective species. Averaging these values over a typical 24-hr period, Nielsen put forward summer rate constants of 1.5% hr^{-1} for tetramethyllead and 5% hr^{-1} for tetraethyllead, and corresponding winter values of 0.3 and 1.1% hr^{-1} [153].

With regard to the other dominant route for TAL breakdown in the atmosphere, the reaction with OH radicals, there appears to be some discrepancy in the literature about the estimated rate constant, particularly for the attack on tetraethyllead; values of 70×10^5 ppm^{-1} hr^{-1}, 27×10^5 ppm^{-1} hr^{-1} and 10.3×10^5 ppm^{-1} hr^{-1} have been reported [216, 226, 225]. Using the median value, which is also the geometric mean of the two outer rate constants [153], and an average hydroxyl radical concentration of 8×10^{-8} ppm and 2.4×10^{-8} ppm in summer and winter respectively [227], one arrives at a depletion rate in summer of 5% hr^{-1} for tetramethyllead and 22% hr^{-1} for tetraethyllead, and corresponding winter values of 2 and 6% hr^{-1}.

The reaction between ozone and R_4Pb in the vapor phase seems to contribute less to the combined sink processes for tetraalkyllead. Based on a representative ozone level of 30 ppb in summer and 20 ppb in winter, decay rates have been suggested of about 0.2% hr^{-1} and 3% hr^{-1} in summer, and 0.14% hr^{-1} and 2% hr^{-1} in winter for tetramethyllead and tetraethyllead respectively [216, 153]. Other homogeneous reactions are apparently not involved to a significant extent.

Heterogeneous reactions have also been investigated, by exposing atmospheric particulate material to gaseous TAL compounds. Neither direct physical adsorption nor surface reactions were claimed to be important as sink mechanism [216].

The combination of these pathways, among which the reaction with OH radicals is clearly predominant, leads to a total atmospheric decomposition rate in summer of 7% hr^{-1} and 30% hr^{-1}, and in winter of 2% hr^{-1} and 9% hr^{-1}, for the respective species. Consequently the half-life of tetraethyllead, which shows a distinctly lower stability than tetramethyllead, is limited to 2–7 hours, depending on the time of the year, whereas for tetramethyllead a half-life of 10–34 hours can be expected. These rather short lifetimes preclude transport over large distances; it can be assumed that TAL from anthropogenic sources would be unable, therefore, to cause significant elevation of atmospheric TAL concentrations in truly remote locations.

At present, efforts are mainly directed to the identification of the degradation products formed in the TAL breakdown process and the determination of their lifetimes. After having established the significance of gas-phase ionic triAL salts in the atmosphere, Hewitt et al. [226] have carried out laboratory experiments to confirm that they are formed by the gas-phase reactions of TAL. After reaction with hydroxyl radicals it was observed that a mixture of tetramethyllead and tetraethyllead produced a suite of lead products; both in the gas and the aerosol phase triAL as well as diAL species were identified, but no redistribution leading to mixed compounds was found to occur. The fraction associated with the aerosol was always small, but a high proportion of the less volatile species (diAL, Pb^{++}) had become adsorbed on the walls of the smog chamber. Assuming that triAL decomposition reactions other than photooxidation by hydroxyl would not be significant in the smog chamber, breakdown rate constants of 2.2×10^5 ppm^{-1} hr^{-1} for trimethyllead and 8.1×10^5 ppm^{-1} hr^{-1} for triethyllead were derived. This would result in a decay rate of 0.5–1.8% hr^{-1} for trimethyllead and 1.9–6.5% hr^{-1} for triethyllead.

As the corresponding half-life of particularly trimethyllead is in the order of days, the trialkyllead species are quite persistent in the atmosphere and may be advected to remote areas distant from vehicular sources. It should be stressed, however, that the limitations of using small smog chamber reactors for simulating atmospheric photochemical kinetic measurements have since long been recognized [226]. Whilst this approach is the only one which allows tetraalkyllead photochemical processes to be elucidated the results obtained must be treated with caution. The reaction sequence which is followed to produce the ionic alkyllead species is at present mainly speculative and needs further confirmation.

Loss of TAL from the atmosphere by washout, rainout, dry deposition and other mechanisms has not explicitly been studied. Taking into account the physico-chemical characteristics of the compounds, these processes are probably of minor significance. As the ionic alkyllead species are well soluble in water, they will be removed from the atmosphere by wet deposition processes..As discussed in a previous section suburban, 'wet' deposition rates for alkyllead at suburban sites amount to 30–60 μg Pb m^{-2} yr^{-1}, according to two measurement campaigns using funnel-in-bottle arrangements [205, 204]. Dry deposition rates are as yet uncertain, due to a lack of reliable values for the deposition velocity. More research is needed to evaluate the impact of these displacement processes.

Aquatic Sink Processes

Tetraalkyllead compounds may disappear from aqueous systems as a result of both photolytical breakdown and volatilization losses. In agreement with the densities of 1.65–1.99 kg dm^{-3} at 20 °C, it has been suggested that TAL in e.g. seawater would lie on the bed in a separate liquid phase, slowly dissolving into the seawater [228]. Some would evaporate to the atmosphere, but most would decompose in situ.

Among the fairly detailed studies which have investigated the decomposition of alkyllead compounds in aqueous solutions, most have been carried out at initial concentrations of $\geq 10^4$ μg Pb l^{-1}, which is many orders of magnitude greater than the amounts encountered in most environmental situations. Particularly for TAL degradation, the outcome of these laboratory experiments should be considered strictly qualitatively, as the observed rates will have been influenced by the solubility and the adsorption on reaction vessels, which may result in heterogeneous reactions atypical for the environment. Nevertheless, it cannot be excluded that such heterogeneous surface reactions are also of consequence in the hydrosphere, for it has been demonstrated that TAL may adsorb on suspended particles in aqueous systems [87, 88].

Generally, the studies agree in that the decomposition of TAL in natural water is a rapid process, light-induced and promoted by various cations [88, 200, 205, 228–230]. Even in the dark, all R_4Pb species decomposed completely within five days when spiked at the low μg l^{-1} level in environmental water [230]. In ultrapure water, some

30% of the tetramethyllead input was recovered after 15 days, but no tetraethyllead remained. The principal degradation product in all the investigations has been found to be trialkyllead, which in turn is decomposed further to inorganic lead. A simulation of the typical breakdown process of tetraethyllead, artificially accelerated by UV irradiation, is shown in Figure 6. Spiking of a mixture of the five TAL species yields, among other products, dimethylethyllead and methyldiethyllead, which are regularly observed in the aqueous environment [230]. These compounds may in addition enter the hydrosphere through wet deposition processes, scavenging the gaseous ionic alkyllead pollutants.

Trialkyllead and dialkyllead species are considerably more stable in the hydrosphere than R_4Pb compounds. In the dark they are particularly persistent and in this way concentrated solutions in ultrapure water can be stored for months without major deterioration. Dilute solutions demonstrate a slow degradation, and after prolonged storage, redistribution reactions may give rise to traces of mixed compounds when methyl- and ethyllead species are kept together [204], though the contrary has also been reported [230]. The decomposition reactions are clearly accelerated by solar radiation [88, 204, 230] and similarly as for the TAL compounds ethyl species have an inferior stability compared to the related methyl-containing products [88, 92, 204, 230]. Fitting the obtained data to a first-order rate expression points to a half-life in the order of 1 day for triethyllead and 10 days for trimethyllead [231].

Interestingly, almost all the data indicate that a simple pathway TAL → TriAL → DiAL → Pb^{++} does not explain the degradation of alkyllead in an aqueous environment. Except in one study [229], dialkylleads have not been detected, or only in trace amounts, as intermediates in the trialkyllead decomposition. The species on

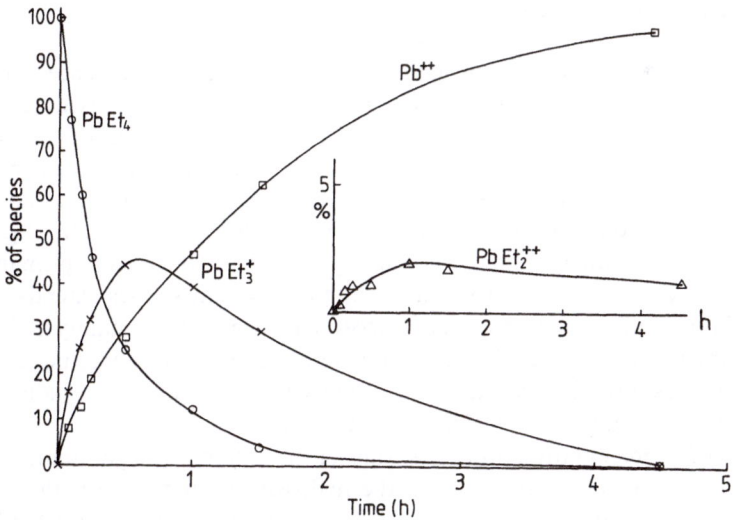

Fig. 6. Simulation of the typical breakdown process of tetraethyllead in deionized water (spiking level 6×10^{-6} M Pb), accelerated through continuous and exclusive irradiation of the quartz erlenmeyer flask with short ultraviolet light (254 nm).
(From reference 210)

itself, however, appear sufficiently stable [204, 230] to exclude a fast conversion to inorganic lead, as suggested in earlier reviews [20, 26]. Whereas it is tempting to assume a direct decomposition of ionic alkyllead species to inorganic lead [230], further experimental verification of e.g. the dismutation reactions which may be involved in the transformations between alkyllead compounds in aqueous systems [88, 165, 224] is required to elucidate the pathway.

There is no clear view either about to what extent adsorption processes constitute a sink for ionic alkyllead species in water. In a solution containing hydrous iron(III)-oxide, no detectable adsorption was encountered for trialkyllead, but dialkyllead was significantly removed [215]. Complete adsorption of all species on silica has also been observed [88]. When samples of natural water are filtered, occasionally a non-negligible part of the analytes appears to be retained with the suspended particles, depending on the type of sample [199, 204, 205]. On the other hand, no appreciable adsorption onto reaction vessel walls seems to occur [205]. Further investigations should reveal the significance of adsorption onto particles and sediments as a sink process for aqueous organolead.

On the fate of alkyllead compounds in soils only sporadic information is available. As in other matrices, possibly spilled TAL compounds are apparently converted quickly to ionic organolead species, which show a much higher plant toxicity and availability than comparable amounts of lead in the form of inorganic salts [232]. The final decomposition to Pb^{++} is assumed to be a slow process [232, 206]. It has been shown that the soluble compounds are leached out easily from soil by water, and that a close relation exists between the level of tetraalkyllead spiked and the water extractable lead; inorganic lead undergoes a rapid and almost complete fixation in the soil [232].

The Biogeochemical Cycle of Lead

To summarize our tentative knowledge about the sources, occurrence and sinks of organolead in the biosphere, biogeochemical cycles have been proposed, schematizing the fate of the pollutants [20, 184, 233, 234]. An updated version of the cycle is represented in Figure 7. Man-made alkyllead enters the environment in the form of tetraalkyllead and probably also ionic alkyllead, emitted into the atmosphere or spilled in the aqueous and terrestrial environment. In both cases the original TAL pollutants have a limited lifetime of a few days at most. Conversion, primarily by reaction with hydroxyl radicals, to trialkyl- and dialkyllead species extends the atmospheric lifetime and allows transport of atmospheric organolead, mainly in the gaseous phase, to more remote areas, before complete breakdown to inorganic lead has occurred. Aqueous tetraalkyllead degrades rapidly to trialkyllead, especially under the influence of light, and finally yields inorganic lead.

Wet and dry precipitation processes transfer the atmospheric pollutants, mainly as ionic tri- and dialkyllead species, into the hydrosphere and deposit them on soils. From natural water and soil the organolead products may get accumulated by plants and animals, in concentrations closely following the environmental exposure, and pass on through the food chain. Apart from the most volatile species which may evaporate, eventually the lead species will gradually degrade or deposit with excreta, dead organisms, or as particulate-bound lead. Sediments and/or biota may in addition

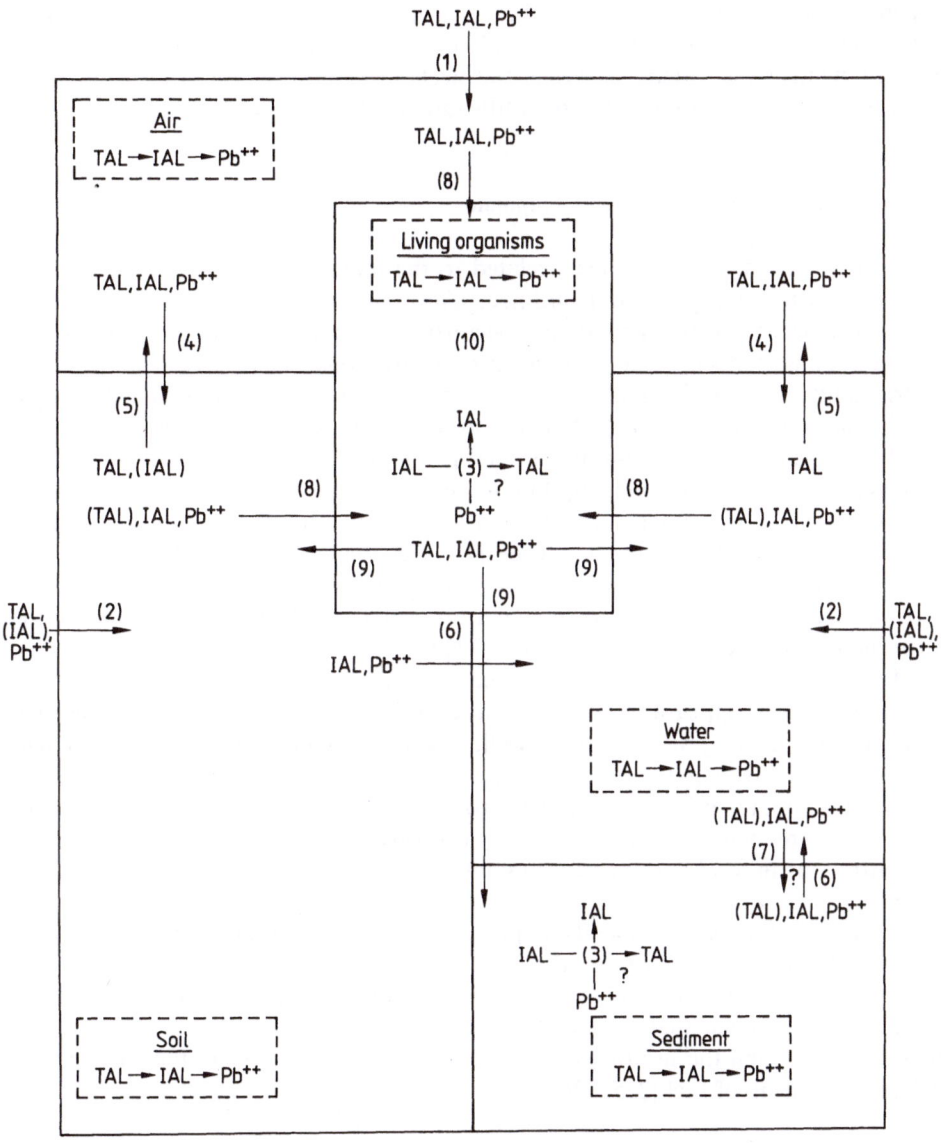

Fig. 7. Simplified biogeochemical cycle of organolead compounds in the environment, showing the flow of the species in and between the different compartments of the biosphere (IAL = ionic alkyllead)

Anthropogenic sources

(1)   vehicle emissions, evaporation

(2)   (accidental) spills, effluents

Natural sources

(3)   (bio)alkylation

Displacement processes

(4)   dry, wet precipitation

(5)   evaporation

(6)   solubilization, remobilization

(7)   deposition of particulate matter

(8)   uptake by living organisms

(9)   deposition from organisms (excreta, dead)

(10)   accumulation through food chain

regenerate alkyllead through a possible environmental alkylation. A probably slow remobilization of settled alkyllead would complete the circle. Particularly the latter stages remain as yet obscure, as well as the mechanism of the various transformations. This prevents at present a sound quantification at the different stages.

Biological Effects of Organolead Compounds

Whereas the human health effects related to the uptake of inorganic lead have extensively been investigated and outlined, almost no assessment has been made to date of the health effects, if any, of non-occupational exposure to organic lead through its use as an antiknock agent. Significant concentrations of inorganic lead are found in foodstuffs and drinking water, and, in general, ingestion represents the major uptake pathway. It is estimated that the average daily uptake of an adult is 23 µg, of which 7.5 µg is derived from the air, 1.1 µg from water, 13 µg from food and the remaining from dust and soil [190]. In a recent large-scale field study in the Northern Italian region of Piedmont it has been attempted to estimate the contribution of gasoline lead to human blood lead levels [13]. For this purpose, an almost complete exchange of the usual gasoline lead, having an isotopic ratio ^{206}Pb to ^{207}Pb around 1.18, was arranged with gasoline having a ratio of 1.04. During the main phase the airborne lead in Turin contained about 90% lead derived from gasoline, and in the rural areas about 60%. The gasoline-related fraction of blood lead of the Turin follow-up subjects appeared to be of the order of 25%, but has been criticized as probably underestimating the true value [190], because the limited size of the experiment may have prevented e.g. the food contribution from adequately being accounted for. Organic lead, on the other hand, appears to be principally present in the gas phase of the atmosphere, hence its uptake pathway is probably dominated by inhalation; the concentrations in food and the resulting intake of organic lead, however, are largely unknown.

The toxic action of organolead compounds was recognized early in the commercial use of TEL, and up to now more than 150 fatal cases of organolead poisoning have been recorded [235]. Potentially exposed occupational groups include tank cleaners, filling station and enclosed car-park staff. Non-occupational acute exposure to TAL may occur through the use of leaded gasoline as a cleaning agent and through the deliberate sniffing of fuel [235, 236].

Uptake and Metabolism

Tetraalkyllead compounds are readily absorbed through the skin, lungs and gastro-intestinal tract. The absorption depends not only on the chemical characteristics but also on the dose, exposure time and animal species. In general, tetraethyllead is better absorbed than tetramethyllead; inorganic lead absorption is substantially less significant [237].

Being a lipophilic substance, tetraethyllead can directly penetrate the skin in lethal quantities [238, 239]. When diluted with gasoline to concentrations around 0.1% (v/v), the absorption becomes slight and inappreciable [239]. Moreover, the possibility of human contact with liquid TAL, other than through occupational exposure or persistent contact with leaded gasoline, is slight, and in Western Europe adequate preventive measures exist [240].

Pulmonary absorption has been investigated by exposing human volunteers to low concentrations of ^{203}Pb-labeled tetraalkyllead [241]. A considerable part of the initially deposited amount, which averaged 40—50% of the inhaled vapor, was subsequently lost by exhalation, due to the reversibility of the transfer from the air in the lungs to the blood. The total amount absorbed in the body for both tetramethyllead and tetraethyllead was 30 to 40% of the inhaled amount of vapor. Extrapolation of these results to the average city air concentration levels would lead to a daily organolead uptake through inhalation of about 500 ng Pb for an adult person. In the experiment, labeled lead in the blood steadily decreased after inhalation ceased, but reappeared in the red cells after 8–20 hours, possibly in the form of triAL.

Gastrointestinal absorption appears also effective, particularly when trialkyllead is orally administered [242]; tetraalkyllead may at least partly decompose to trialkyllead under the influence of gastric hydrochloric acid [145].

Animal experiments set up to reveal the metabolism of organic lead have indicated that the toxicity of tetraalkyllead is caused by trialkyllead formed through degradation processes in the body, and mainly in the liver [128]. The necessary metabolism and the succeeding redistribution of trialkyllead to target sites, primarily the brain, takes some time and may partly explain the relatively long period from administration of TAL to the appearance of neurological symptoms [237]. The metabolic conversion is an oxidative dealkylation in the liver microsomes, involving the monooxygenase system [128, 243, 244]. Once formed, trialkyllead compounds seem to be fairly stable in biological tissues, but further degradation to dialkyllead and inorganic lead may take place, as supported by e.g. urine determinations [222, 223, 237].

The organic lead absorbed is transported by the blood to other parts of the body and accumulates mainly in the liver, and to a lesser extent also in kidney and brain [e.g. 128, 245]. The brain is considered the critical organ in organolead intoxication. Triethyllead shows an interesting specificity in its inhibitory action on glucose oxidation by brain slices: it does not inhibit this process in kidney or liver slices, nor does it inhibit the oxidation of substrates other than glucose by brain tissue [246, 247]. Fundamental differences thus exist between the toxic mechanisms of organic and inorganic lead, since tetraethyllead and lead acetate have no effect on the lactate metabolism [128, 247].

Organic lead is mostly eliminated by excretion as inorganic lead in the faeces, with less than one third found in the urine [239, 237]; in the latter case dominantly dialkyllead species are found. Whereas the blood lead level is not considered a reliable index of long-term organolead retention, due to the short residence time of lead in the blood, the commonly used alternative index, lead in the urine, is also of questionable value. A lack of correlation between urinary levels and effects on the central nervous system has been demonstrated, and the excretion of lead in the urine is relatively low and variable [237].

Toxicology

A hypothetical mechanism for the selective neurotoxic action of triethyllead and trimethyllead, as suggested recently by Cremer [248], implies that they bring about an increase in neuronal excitability, probably by altering Cl^- distribution across neuronal membranes. This increase in neuronal activity, reflected in behavioural symptoms, triggers an increase in glucose utilization. The increased metabolic rate of glucose exceeds the maximal capacity for pyruvate oxidation and coupled ATP production, resulting in accumulation of pyruvic and lactic acid, which causes a localized acidosis. In addition, both alkyllead species may directly act on mitochondria to inhibit the entry of substrates and the synthesis of ATP [249]; the binding sites are possibly thiol groups.

As plants lack any structures comparable to mammalian nervous tissue, the main target of organolead attack, their response to intoxication with organic lead compounds may greatly differ from that of mammals. In accordance with diverse findings in mammalian organisms, R_4Pb must be decomposed into the R_3Pb^+ form to resume toxicity to plants. Though up to now no evidence exists indicating metabolic dealkylation of R_4Pb by any plant species, physico-chemical mechanisms in the environment provide effective abiotic decomposition of R_4Pb into the highly toxic ionic derivatives. Extensive studies by Röderer [250–253] have demonstrated that, in the dark, TAL has no toxic action at all on the unicellular alga Poterioochromonas malhamensis, but upon illumination triAL inhibits growth, mitosis, cytokinesis and lorica formation, and causes the formation of giant multinucleate cells. These observations point to an interaction with microtubules, which was subsequently confirmed in plant but also in mammalian cells [254–256]. The triethyllead ion was found to interact with 2 out of 18 thiol groups present in tubulin dimers, thus inhibiting the microtubule assembly [257]. The high tubulin content of cells from the nervous system has been proposed to be the reason for the selective neurotoxicity, as particularly the working of these cells would depend on the integrity of the microtubules, which play a basic role in shaping of cells and in axoplasmic transport processes [258].

Whereas TAL in itself appears not to represent any toxic hazard for algae, higher aquatic species show a different response. The water-soluble triAL compounds may have difficulty in penetrating the gill membrane, whereas the hydrophobic TAL compounds are readily absorbed and metabolized [167]. This leads to the observation that trialkyllead compounds are apparently less toxic than tetraalkyllead, and explains the higher toxicity of tetraethyllead compared to tetramethyllead [259].

The lethal dose for man is not known, but can be estimated from the observed LD_{50} for mammals (i.e. the quantity of substance per unit body weight given as a single dose required to kill half the exposed population within 14 days). This has been found to be <36 mg Pb kg^{-1} for trimethyllead chloride, 80 mg Pb kg^{-1} for tetramethyllead, 20 mg Pb kg^{-1} for triethyllead chloride and 15 mg Pb kg^{-1} for tetraethyllead, when oral administration is applied [145], leading to a lethal dose for an adult person of about 0.25 g tetraethyllead and more than 1 g tetramethyllead. The greater toxicity of the former compound is supposed to reflect its faster dealkylation in vivo.

The symptoms of alkyllead poisoning in man are in the early period rather vague and nonspecific: fatigue, headache, vomiting, diarrhea and particularly insomnia may occur. Later, signs of involvement of the central nervous system develop, like tremor, and vegetative disturbances (hypothermia, hypotonia) [235]. No effective treatment has been recommended; the beneficial effect on organolead poisoning of typical chelating agents used for Pb^{++} has to be considered doubtful [235].

As a thorough discussion of the biological effects of organolead is out of the scope of this environmental review, the interested reader is further referred to several excellent surveys which have recently been published [21, 260, 261].

References

1. Nriagu JO (1978) Properties and the biogeochemical cycle of lead. In: Nriagu JO (ed) The biogeochemistry of lead in the environment, part A, Elsevier, Amsterdam, p 1
2. Waldron HA, Stoffen D (1974) Sub-clinical lead poisoning. Academic, New York
3. Gilfillan SC (1965) J. Occup. Med. 7: 53
4. Midgley TJr, Boyd TA (1922) Ind. Eng. Chem. 14: 894
5. Hewitt CN, Harrison RM (1986) Organolead compounds in the environment. In: Craig PJ (ed) Organometallic compounds in the environment, Longman Harlow, p 160
6. Robinson IM (1978) Lead as a factor in the world economy. In: Nriagu JO (ed) The biogeochemistry of lead in the environment, part A, Elsevier, Amsterdam, p 99
7. Löwig C (1852) Chem. Zentr. 1852: 575
8. Buckton GB (1859) Proc. Roy. Soc. IX: 685
9. Pfeiffer P, Truskier P (1904) Chem. Ber. 37: 1125
10. Gruttner G, Krause E (1916) Chem. Ber. 49: 1125
11. Murozumi M, Chow TJ, Patterson C (1969) Geochim. Cosmochim. Acta 33: 1247
12. Bruland KW, Bertine K, Koide M, Goldberg ED (1974) Environ. Sci. Technol. 8: 425
13. Facchetti S, Geiss F, Gaglione P, Colombo A, Garibaldi G, Spallanzani G, Gilli G (1982) Isotopic Lead Experiment Status Report, Commission of the European Communities, Report EUR 8352 EN
14. Wolff EW, Peel DA (1985) Nature 313: 535
15. Trefry JH, Metz S, Trocine RP, Nelsen TA (1985) Science 230: 439
16. Eisenreich SJ, Metzer NA, Urban NR, Robbins JA (1986) Environ. Sci. Technol. 20: 171
17. Jensen RA, Laxen DPH (1987) Sci. Tot. Environ. 59: 1
18. Dickson D (1985) Science 228: 159
19. Larbey RJ (1986) Legislation on petrol lead content and vehicle exhaust emissions in Western Europe, its cost and other implications. In: Lester JN, Perry R, Steritt RM (eds) Proc. Int. Conf. Chemicals in the Environment, Lisbon, Selper, London, p 843
20. De Jonghe WRA, Adams FC (1986) Biogeochemical cycling of organic lead compounds. In: Nriagu JO, Davidson CI (eds) Toxic metals in the atmosphere, J. Wiley, New York, p 561
21. Grandjean P (ed) (1984) Biological effects of organolead compounds, CRC Press, Boca Raton, FL
22. Calingaert G (1926) Chem. Rev. 2: 43
23. Krause E, von Grosse A (1937) Die Chemie der metallorganischen Verbindungen. Borntraeger, Berlin, p 372
24. Leeper RW, Summers L, Gilman H (1954) Chem. Rev. 54: 101
25. Milde RL, Beatty HA (1959) Adv. Chem. Ser. 23: 306
26. Shapiro H, Frey FW (1968) The organic compounds of lead. J. Wiley, New York
27. Frey FW, Shapiro H (1971) Top. Curr. Chem. 16: 243
28. Gilroy KM, Price SJ, Webster NJ (1972) Can. J. Chem. 50: 2639
29. Lewis GL, Oesper PF, Smyth CP (1940) J. Am. Chem. Soc. 62: 3243
30. Calingaert G, Beatty HA (1939) J. Am. Chem. Soc. 61: 2748
31. Calingaert G, Beatty HA (1943) The redistribution reaction. In: Gilman H (ed) Organic chemistry, an advanced treatise, vol II, 2nd edn, Wiley, New York, p 1806
32. de Vos D, Wolters J (1980) Rev. Silicon, Germanium, Tin and Lead Compounds 4: 209
33. Blais JS, Marshall WD (1987) Appl. Organometall. Chem. (in press)
34. Chobert G, Devaud M (1978) J. Organometall. Chem. 153: C23
35. Tiravanti G, Boari G (1979) Environ. Sci. Technol. 13: 849
36. Harrison GF (1980) The Cavtat incident. In: Branica M, Konrad Z (eds) Lead in the marine environment, Pergamon, Oxford, p 305

37. Snyder LJ, Henderson SR (1961) Anal. Chem. *33*: 1175
38. Linch AL, Davis RB, Stalzer RF, Anzilotti WF (1964) Am. Ind. Hyg. Assoc. J. *25*: 81
39. Moss R, Browett EV (1966) Analyst *91*: 428
40. Purdue LJ, Enrione RE, Thompson RJ, Bonfield BA (1973) Anal. Chem. *45*: 527
41. Walker AO (1975) U.S. patent 3 870 469
42. Cope RF, Pancamo BP, Rinehart WE, ter Haar GL (1979) J. Am. Ind. Hyg. Assoc. *40*: 372
43. Mitchell WJ, Midgett MR (1979) J. Air Pollut. Contr. Assoc. *29*: 959
44. Seeley JL, Skogerboe RK (1974) Anal. Chem. *46*: 415
45. Robinson JW, Rhodes L, Wolcott DK (1975) Anal. Chim. Acta *78*: 474
46. Sawicki CR (1975) Seminar summary: sampling and analysis of the various forms of atmospheric lead, report no EPA-650/2-75-003, NERC North Carolina
47. Harrison RM, Perry R (1977) Atmos. Environ. *11*: 847
48. Robinson JW (1978) Atmos. Environ. *12*: 1247
49. De Jonghe WRA, Adams FC (1982) Talanta *29*: 1057
50. Hancock S, Slater A (1975) Analyst *100*: 422
51. Birch J, Harrison RM, Laxen DPH (1980) Sci. Tot. Environ. *14*: 31
52. De Jonghe WRA, Adams FC (1980) Atmos. Environ. *14*: 1177
53. Gibson MJ, Farmer JG (1981) Environ. Technol. Lett. *2*: 521
54. Febo A, Di Palo V, Possanzini M (1986) Sci. Tot. Environ. *48*: 187
55. Snyder LJ (1967) Anal. Chem. *39*: 591
56. Barker AJ, Mulligan RL (1978) A thermoanalytical investigation of TEL laden activated carbons. In: Proc. 5th Int. Carbon Conf., London, p 171
57. Birnie SE, Noden FG (1980) Analyst *105*: 110
58. Røyset O, Thomassen Y (1986) Anal. Chim. Acta *188*: 247
59. Rohbock E, Müller J (1979) Microchim. Acta *1*: 423
60. Radziuk B, Thomassen Y, Van Loon JC, Chau YK (1979) Anal. Chim. Acta *105*: 255
61. De Jonghe WRA, Chakraborti D, Adams FC (1981) Atmos. Environ. *15*: 421
62. Rohbock E, Georgii H-W, Müller J (1981) Atmos. Environ. *15*: 422
63. Laveskog A (1970) A method for determination of tetramethyllead (TML) and tetraethyllead (TEL) in air. In: Proc. 2nd Int. Clean Air Congress, Washington DC, p 549
64. Harrison RM, Perry R, Slater DH (1974) Atmos. Environ. *8*: 1187
65. Harrison RM, Perry R, Slater DH (1974) The contribution of organic lead compounds to total lead levels in urban atmospheres. In: Proc. Int. Symp. Recent advances in assessment of health effects of environmental pollution, Paris, p 1783
66. Chau YK, Wong PTS, Saitoh H (1976) J. Chromatog. Science *14*: 162
67. Robinson JW, Kiesel EL, Goodbread JP, Bliss R, Marshall R (1977) Anal. Chim. Acta *92*: 321
68. Reamer DC, Zoller WH, O'Haver TC (1978) Anal. Chem. *50*: 1449
69. De Jonghe WRA, Chakraborti D, Adams FC (1980) Anal. Chem. *52*: 1974
70. Nielsen T, Egsgaard H, Larsen E, Schroll G (1981) Anal. Chim. Acta *124*: 1
71. Jiang SG, Chakraborti D, De Jonghe W, Adams F (1981) Z. Anal. Chem. *305*: 177
72. De Jonghe WRA (1987) Personal communication
73. Cantuti V, Cartoni GP (1968) J. Chromatog. *32*: 641
74. Coker DT (1978) Ann. Occup. Hyg. *21*: 33
75. Hewitt CN, Harrison RM (1985) Anal. Chim. Acta *167*: 277
76. Chau YK, Wong PTS, Goulden PD (1976) Anal. Chim. Acta *85*: 421
77. Hewitt CN, Harrison RM (1985) Total speciation of gas-phase alkyllead in the atmosphere. In: Lekkas TD (ed) Proc. Int. Conf. Heavy Metals in the Environment, Athens, CEP Consult. Edinburgh, p 171
78. Harrison RM, Radojevic M, Hewitt CN (1985) Sci. Tot. Environ. *44*: 235
79. Hewitt CN, Harrison RM, Radojevic M (1986) Anal. Chim. Acta *188*: 229
80. Hewitt CN, Harrison RM (1987) Environ. Sci. Technol. *21*: 260
81. De Jonghe WRA, Chakraborti D, Adams FC (1981) Environ. Sci. Technol. *15*: 1217
82. Potter HR, Jarvie AWP, Markall RN (1977) Wat. Pollut. Control *76*: 123
83. Chau YK, Wong PTS, Bengert GA, Kramar O (1979) Anal. Chem. *51*: 186
84. Noden FG (1980) The determination of tetraalkyllead compounds and their degradation products in natural water. In: Branica M, Konrad Z (eds) Lead in the marine environment. Pergamon, Oxford, p 83

85. Van Cleuvenbergen R, Chakraborti D, Adams F (1987) Spciation of ionic organolead compounds in the biosphere. In: Angeletti G, Restelli G (eds), C. E. C., Physico-chemical behaviour of atmospheric pollutants, D. Reidel, p 571
86. Cruz RB, Lorouso C, George S, Thomassen Y, Kinrade JD, Butler LRP, Lye J, Van Loon JC (1980) Spectrochim. Acta *35B*: 775
87. Robinson JW, Rhodes IAL (1980) J. Environ. Sci. Health *A15*: 201
88. Jarvie AWP, Markall RN, Potter HR (1981) Environ. Res. *25*: 241
89. Diehl KH, Rosopulo A, Kreuzer W (1983) Fres. Z. Anal. Chem. *314*: 755
90. Aneva Z (1985) Fres. Z. Anal. Chem. *321*: 680
91. Bolanowska W (1967) Chem. Analityczna *12*: 121
92. De Jonghe WRA, Van Mol WE, Adams FC (1983) Anal. Chem. *55*: 1050
93. Pilloni G, Plazzogna G (1966) Anal. Chim. Acta *35*: 325
94. Schmidt U, Huber F (1978) Anal. Chim. Acta *98*: 147
95. Blaszkewicz M, Baumhoer G, Neidhart B (1984) Fres. Z. Anal. Chem. *317*: 221
96. Blaszkewicz M, Baumhoer G, Neidhart B (1987) Int. J. Environ. Anal. Chem. *28*: 207
97. Henderson SR, Snyder LJ (1961) Anal. Chem. *33*: 1172
98. Crompton TR (1974) Chemical analysis of organometallic compounds, vol. 3, Academic, London, p 101
99. Barker AJ, Ellis SRM, Clarke AB (1976) Chem. Ser. ACS *155*: 381
100. Chau YK, Wong PTS, Kramar O (1983) Anal. Chim. Acta *146*: 211
101. Estes SA, Uden PC, Barnes RM (1982) Anal. Chem. *54*: 2402
102. Forsyth DS, Marshall WD (1983) Anal. Chem. *55*: 2132
103. Radojevic M, Allen A, Rapsomanikis S, Harrison RM (1986) Anal. Chem. *58*: 658
104. Chakraborti D, De Jonghe WRA, Van Mol WE, Van Cleuvenbergen RJA, Adams FC (1984) Anal. Chem. *56*: 2692
105. Van Cleuvenbergen RJA, Van Mol WE, Adams FC (1988) J. Anal. At. Spectrom. *3*: 169
106. Baussand P, Foster P, Besson J, Laverlochere J, Meinhrat AO (1985) Analusis *13*: 53
107. Foster P, Laffond M, Perraud R, Baussand P, Jacob V (1987) Int. J. Environ. Anal. Chem. *28*: 105
108. D'Ulivo A, Fuoco R, Papoff P (1986) Talanta *33*: 401
109. Rapsomanikis S, Donard OFX, Weber JH (1986) Anal. Chem. *58*: 35
110. Hodges DJ, Noden FG (1979) The determination of alkyllead species in natural waters by polarographic techniques. In: Proc. Int. Conf. Heavy Metals in the Environment, London, CEP Consult. Edinburgh, p 408
111. Colombini MP, Corbini G, Fuoco R, Papoff P (1981) Ann. Chim. *71*: 609
112. Colombini MP, Fuoco R, Papoff P (1984) Sci. Tot. Environ. *37*: 61
113. Bond AM, Bradbury JR, Howell GN, Hudson HA, Hanna PJ, Strother S (1983) J. Electroanal. Chem. *154*: 217
114. Bond AM, Bradbury JR, Hanna PJ, Howell GN, Hudson HA, Strother S, O'Connor MJ (1984) Anal. Chem. *56*: 2392
115. Edwards HW, Rosenvold RJ (1974) Uptake of tetraethyllead vapor by atmospheric dust components. In: Proc. 2nd Ann. NSF-RANN Trace Contaminants Conf., California, p 59
116. Harrison RM (1976) J. Environ. Sci. Health *A11*: 417
117. Sirota GR, Uthe JF (1977) Anal. Chem. *49*: 823
118. Jennen A, Delafortrie A, Verdoodt D, Jacobs T, Dourte P (1984) Landbouwtijdschrift *37*: 1011 (in Dutch)
119. Hayakawa K (1971) Jap. J. Hyg. *26*: 377
120. Andersson K, Nilsson CA, Nygren O (1984) Scand. J. Work Environ. Health *10*: 51
121. Beccaria AM, Mor ED, Poggi G (1978) Ann. Chim. *68*: 607
122. Mor ED, Beccaria AM (1980) A dehydratation method to prevent loss of trace elements in biological samples. In: Branica M, Konrad Z (eds) Lead in the marine environment, Pergamon, Oxford, p 53
123. Stevens CD, Feldhake CJ, Kehoe RA (1960) J. Pharmacol. Exp. Ther. *128*: 90
124. Bolanowska W, Piotrowski J, Garczynski H (1967) Archiv für Toxikologie *22*: 278
125. Birnie SE, Hodges DJ (1981) Environ. Technol. Lett. *2*: 433
126. Yamaushi H, Arai F, Yamamura Y (1981) Ind. Health *19*: 115
127. Orren DK, Caldwell-Kenkel JC, Mushak P (1985) J. Anal. Toxicol. *9*: 258
128. Cremer JE (1959) Brit. J. Industr. Med. *16*: 191

129. Chau YK, Wong PTS, Bengert GA, Dunn JL (1984) Anal. Chem. *56*: 271
130. Forsyth DS, Marshall WD (1986) Environ. Sci. Technol. *20*: 1033
131. Blais JS, Marshall WD (1986) J. Environ. Qual. *15*: 255
132. Radojevic M, Harrison RM (1986) Environ. Technol. Lett. *7*: 525
133. Nygren O, Nilsson C-A (1987) J. Anal. At. Spectrom. *2*: 805
134. Chakraborti D, Van Cleuvenbergen R, Adams F (1986) Determination of ionic alkyllead compounds in road runoff water and road dust by butylation and gas chromatography-atomic absorption spectrometry. In: Lester JN, Perry R, Steritt RM (eds) Proc. Int. Conf. Chemicals in the Environment. Lisbon, Selper, London, p 298
135. Chakraborti D, Van Cleuvenbergen R, Adams F (1987) Int. J. Environ. Anal. Chem. *30*: 233
136. Chakraborti D, Van Cleuvenbergen RJA, Adams FC (1989) Hydrobiologia (in press)
137. Barry PSI (1975) Postgrad. Med. J. *51*: 783
138. Ter Haar GL, Bayard MA (1971) Nature *232*: 553
139. Hirschler DA, Gilbert LF (1964) Arch. Environ. Health *8*: 297
140. Biggins PDE, Harrison RM (1979) Environ. Sci. Technol. *13*: 558
141. Huntzicker JJ, Friedlander SK, Davidson CI (1975) Environ. Sci. Technol. *9*: 448
142. Hewitt CN, Rashed MB (1988) Appl. Organometall. Chem. *2*: 95
143. Laveskog A (1972) Organolead compounds in auto exhaust and street air. TPM-BIL-64, AB Atomenergic, Stockholm
144. Rohbock E, Georgii H-W, Müller J (1980) Atmos. Environ. *14*: 89
145. Grandjean P, Nielsen T (1979) Residue Rev. *72*: 97
146. Kehoe RA, Cholak J, Spence JA, Hancock W (1963) Arch. Environ. Health *6*: 239
147. Pattenden NJ, Branson JR (1987) Atmos. Environ. *21*: 2481
148. Page RA, Cawse PA, Baker SJ (1988) Sci. Tot. Environ. *68*: 71
149. Laveskog A (1984) Gasoline additives: past, present and future. In: Grandjean P (ed) Biological effects of organolead compounds. CRC, Boca Raton, FL, p 5
150. Wong PTS, Chau YK, Luxon PL (1975) Nature *253*: 263
151. Harrison RM, Laxen DPH (1978) Nature *275*: 738
152. Hewitt CN, de Mora SJ, Harrison RM (1984) Marine Chem. *15*: 189
153. Nielsen T (1984) Atmospheric occurrence of organolead compounds. In: Grandjean P (ed) Biological effects of organolead compounds, CRC Press, Boca Raton, FL, p 43
154. Chau YK, Wong PTS (1986) Occurrence of molecular and ionic alkyllead compounds in environmental samples. In: Lester JN, Perry R, Steritt RM (eds) Proc. Int. Conf. Chemicals in the Environment, Lisbon, Selper, London, p 77
155. Jiang S, Ma C, Liu H, Ge J, Li M, Adams FC, Winchester JW (1984) Atmos. Environ. *18*: 2553
156. Harrison RM, Hewitt CN (1986) Atmos. Environ. *20*: 413
157. Jarvie AWP, Markall RN, Potter HR (1975) Nature *255*: 217
158. Dumas JP, Pazdernik L, Belloncik S (1977) Methylation of lead in the aquatic environment. In: Proc. 12th Can. Symp. Water Pollut. Res. *12*: 91
159. Chau YK, Wong PTS (1980) Biotransformation and toxicity of lead in the aquatic environment. In: Branica M, Konrad Z (eds) Lead in the marine environment, Pergamon, Oxford, p 225
160. Thompson JAJ, Crerar JA (1980) Mar. Pollut. Bull. *11*: 251
161. Craig PJ (1980) Environ. Technol. Lett. *1*: 17
162. Reisinger K, Stoeppler M, Nurnberg HW (1981) Nature *291*: 228
163. Jarvie AWP, Whitmore AP, Markall RN, Potter HR (1983) Environ. Pollut. Ser. B *6*: 81
164. Schmidt U, Huber F (1976) Nature *259*: 157
165. Huber F, Schmidt U, Kirchman H (1978) ACS Symp Ser. *82*: 65
166. Ridley WP, Dizikes LJ, Wood JM (1977) Science *197*: 329
167. Wood JM (1980) Lead in the marine environment — some biochemical considerations. In: Branica M, Konrad Z (eds) Lead in the marine environment, Pergamon, Oxford, p 299
168. Rapsomanikis S, Ciejka JJ, Weber JH (1984) Inorg. Chim. Acta *89*: 179
169. Taylor RT, Hanna ML (1976) J. Environ. Sci. Health — Environ. Sci. Eng. *A11*: 201
170. Agnes G, Bendle S, Hill HAO, Williams FR, Williams RJP (1971) Chem. Commun. *1971*: 850
171. Wood JM (1974) Science *183*: 1049
172. Rhode SF, Weber JH (1984) Environ. Technol. Lett. *5*: 63
173. Ahmad I, Chau YK, Wong PTS, Carty AJ, Taylor L (1980) Nature *287*: 716.
174. Snyder LJ, Bentz JM (1982) Nature *296*: 228

175. Jarvie AWP, Whitmore AP (1981) Environ. Technol. Lett. *2*: 197
176. Kehoe RA, Cholak J, Mc Ilhinney JG, Lofquist GA, Sterling TD (1963) Arch. Environ. Health *6*: 81
177. Cholak J (1964) Arch. Environ. Health *8*: 314
178. Colwill DM, Hickman AJ (1973) Transport and Road Research Laboratory Report LR 545, Dept. Environ. Berkshire (England)
179. Allvin B, Berg S (1977) Analysis of tetraalkyllead in street air, SNV PM 907, Swedish Environmental Protection Agency, Stockholm
180. Rohbock E, Schmitt G (1981) Environ. Technol. Lett. *2*: 263
181. Harrison RM, Laxen DPH, Birch J (1979) Tetraalkyllead in air: sources, sinks and concentrations. In: Proc. Int. Conf. Heavy Metals in the Environment, London, p 257
182. Edwards HW, Rosenvold RJ, Wheat HG (1975) Sorption of organic lead vapor on atmospheric dust particles. In: Hemphill DD (ed) Trace substances in Environmental Health, Univ. Missouri Press, p 197
183. Harrison RM, Laxen DPH (1977) Atmos. Environ. *11*: 201
184. Harrison RM, Hewitt CN, Radojevic M (1985) Environmental pathways of alkyllead compounds. In: Lekkas TD (ed) Proc. Int. Conf. Heavy Metals in the Environment, Athens, CEP Consult. Edinburgh, p 82
185. Allen AG, Radojevic M, Harrison RM (1988) Environ. Sci. Technol. *22*: 517
186. De Jonghe WRA (1983) Ph. D. Thesis, University of Antwerp, Department of Chemistry, p 87 and p 122
187. Cautreels W, Van Cauwenberghe K (1978) Atmos. Environ. *12*: 1133
188. Harrison RM, Radojevic M (1985) Environ. Technol. Lett. *6*: 129
189. Rifkin EB, Walcott C (1956) Ind. Eng. Chem. *48*: 1532
190. Fergusson JE (1986) Sci. Tot. Environ. *50*: 1
191. Radojevic M, Allen A, Hewitt CN (1987) Complete physico-chemical speciation of deposited and atmospheric alkyllead. In: Proc. Int. Conf. Heavy Metals in the Environment, New Orleans, CEP Consult. Edinburgh, p 256
192. Tiravanti G, Rozzi A, Dall'Aglio M, Delaney W, Dadone A (1980) Prog. Wat. Tech. *12*: 49
193. Malizia E, Stacchini A, Costantini S, Baldini M, Giordano R (1978) Riv. Tossicol. Sper. Clin. *8*: 431
194. Radojevic M, Harrison RM (1986) Environ. Technol. Lett. *7*: 519
195. Wong PTS, Chau YK, Yaromich J, Hodson P, Whittle M (1988) Alkyllead contaminations in the St. Lawrence River and St. Clair River (1981–1987). Can. Tech. Rep. Fish. Aquat. Sci. 1602
196. Chau YK, Wong PTS, Kramar O, Bengert GA, Cruz RB, Kinrade JO, Lye J, Van Loon JC (1980) Bull. Environ. Cont. Toxicol. *24*: 265
197. Chau YK, Wong PTS, Bengert GA, Dunn JL, Glen B (1985) J. Great Lakes Res. *11*: 313
198. Harrison RM, Radojevic M, Wilson SJ (1986) Sci. Tot. Environ. *50*: 129
199. Van Cleuvenbergen RJA, Chakraborti D, Adams FC (1986) Environ. Sci. Technol. *20*: 589
200. Grove JR (1980) Investigations into the formation and behaviour of aqueous solutions of lead alkyls. In: Branica M, Konrad Z (eds) Lead in the marine environment, Pergamon, Oxford, p 45
201. De Jonghe WRA, Van Mol WE, Adams FC (1983) Occurrence and fate of organic lead compounds in the hydrosphere. In: Proc. Int. Conf. Heavy Metals in the Environment, Heidelberg, p 166
202. Van Cleuvenbergen R, Chakraborti D, Van Mol W, Adams F (1985) Ionic alkyllead species in atmospheric fall-out and surface water. In: Lekkas TD (ed) Proc. Int. Conf. Heavy Metals in the Environment, Athens, CEP Consult. Edinburgh, p 153
203. Faulstich H, Stournaras C (1985) Nature *317*: 714
204. Van Cleuvenbergen R, Adams F (1988) (unpublished results)
205. Radojevic M, Harrison RM (1987) Atmos. Environ. *21*: 2403
206. Chakraborti D, Dirkx W, Van Cleuvenvergen RJA, Adams FC (1989) Sci. Tot. Environ.
207. Faulstich H, Stournaras C, Endres KP (1987) Experientia *43*: 115
208. Unsworth MH, Harrison RM (1985) Nature *317*: 674
209. Röderer G (1985) Mineralöl — Mineralölrundsch. *33*: 169
210. Van Cleuvenbergen R (1986) Chemie Magazine *12*: 27 (in Dutch)
211. Faulstich H, Stournaras C (1986) Nature *319*: 17
212. Harrison RM (1986) personal communication

213. Backhaus B, Backhaus R (1986) Sci. Tot. Environ. *50*: 223
214. Van Cleuvenbergen RJA, Chakraborti D, Adams FC (1986) Anal. Chim. Acta *182*: 239
215. Riley JP, Towner JV (1984) Mar. Poll. Bull. *15*: 153
216. Harrison RM, Laxen DPH (1978) Environ. Sci. Technol. *12*: 1384
217. Bull KR, Every WJ, Freestone P, Hall JR, Osborn D, Cooke AS, Stowe T (1983) Environ. Poll. Ser. *A31*: 239
218. Head PC, d'Arcy BJ, Osbaldeston PJ (1980) The Mersey estuary bird mortality autumn-winter 1979 preliminary report, as cited in ref. 5
219. Johnson MS, Pluck H, Hutton M, Moore G (1982) Arch. Environ. Contam. Toxicol. *11*: 761
220. Krishnan K, Forsyth DS, Marshall WD, Hatch WI (1987) (submitted for publication)
221. Nielsen T, Jensen KA, Grandjean P (1978) Nature *274*: 602
222. Yamamura Y, Arai F, Yamaushi H (1981) Ind. Health *19*: 125
223. Turlakiewicz Z, Jakubowski M, Chmielnicka J (1985) Br. J. Ind. Med. *42*: 63
224. Haupt HJ, Huber F, Gmehling J (1972) Z. Anorg. Allg. Chem. *390*: 31
225. Nielsen OJ, Nielsen T, Pagsberg P (1982) Direct spectrokinetic investigation of the reactivity of OH with tetraalkyllead compounds in the gas phase. Estimate of lifetimes of TAL compounds in ambient air, Risø Report R-463, Risø National Laboratory, Roskilde, Denmark
226. Hewitt CN, Harrison RM (1986) Environ. Sci. Technol. *20*: 797
227. Hewitt CN, Harrison RM (1985) Atmos. Environ. *19*: 545
228. Robinson JW, Kiesel EL, Rhodes IAL (1979) J. Environ. Sci. Health *A14*: 65
229. Charlou JL, Caprais MP, Blanchard G, Martin G (1982) Environ. Technol. Lett. *3*: 415
230. Harrison RM, Hewitt CN, Radojevic M (1986) Decomposition chemistry of organolead compounds in the environment. In: Lester JN, Perry R, Steritt RM (eds) Proc. Int. Conf. Chemicals in the Environment, Lisbon, Selper, London, p 110
231. Radojevic M, Harrison RM (1987) Sci. Tot. Environ. *59*: 157
232. Diehl KH, Rosopulo A, Kreuzer W, Judel GK (1983) Z. Pflanzenernaehr. Bodenk. *146*: 551
233. Röderer G (1981) Fate and toxicity of tetraalkyllead and its derivatives in aquatic environments. In: Proc. Int. Conf. Heavy Metals in the Environment, Amsterdam, p 250
234. De Jonghe W, Jiang S, Adams F (1981) Alkyllead compounds in the environment. In: Proc. Int. Conf. on Environmental Pollution, Thessaloniki, p 183
235. Grandjean P (1984) Organolead exposures and intoxications. In: Grandjean P (ed) Biological effects of organolead compounds, CRC Press, Boca Raton, FL, p 227
236. Keenlyside RA (1984) The gasoline-sniffing syndrome. In: Grandjean P (ed) Biological effects of organolead compounds, CRC Press, Boca Raton, FL, p 219
237. Jensen AA (1984) Metabolism and toxicokinetics. In: Grandjean P (ed) Biological effects of organolead compounds, CRC Press, Boca Raton, FL, p 97
238. Laug EP, Kunze FM (1948) J. Ind. Hyg. *30*: 256
239. Kehoe RA, Thamann F (1931) Am. J. Hyg. *13*: 478
240. Gething J, Oxley GR (1984) Preventive measures in the occupational setting. In: Grandjean P (ed) Biological effects of organolead compounds, CRC Press, Boca Raton, FL, p 243
241. Heard MJ, Wells AC, Newton D, Chamberlain AC (1979) Human uptake and metabolism of tetraethyl and tetramethyl lead vapor labelled with [203]Pb. In: Proc. Int. Conf. Heavy Metals in the Environment, London, p 103
242. Williams B, Dring LG, Williams RT (1972) Biochem. J. *127*: 24
243. Casida JE, Kimmel EC, Holm B, Widmark G (1971) Acta Chem. Scand. *25*: 1497
244. Ferreira da Silva D, Schröder U, Diehl H (1985) Metabolism of tetraorganolead compounds by rat-liver microsomal mono-oxygenase, Report No 22, Fachberichte Physik, Universität Bremen
245. Osborn D, Every WJ, Bull KR (1983) Environ. Pollut. Ser. A *31*: 261
246. Aldridge WN, Cremer JE, Threlfall CJ (1962) Biochem. Pharmacol. *11*: 835
247. Cremer JE (1965) Rev. Hyg. Profess. *17*: 15
248. Cremer JE (1984) Possible mechanisms for the selective neurotoxicity. In: Grandjean P (ed) Biological effects of organolead compounds, CRC Press, Boca Raton, FL, p 207
249. Aldridge WN (1984) Effects on mitochondria and other enzyme systems. In: Grandjean P (ed) Biological effects of organolead compounds, CRC Press, Boca Raton, FL, p 137
250. Röderer G (1980) Environ. Res. *23*: 371
251. Röderer G (1981) Environ. Res. *25*: 361

252. Röderer G (1982) J. Environ. Sci. Health *A17*: 1
253. Röderer G (1986) Environ. Res. *39*: 205
254. Röderer G (1984) Selective interaction of triethyllead with microtubules from mammalian and nonmammalian cells. In: Hemphill DD (ed) Trace Substances in Environmental Health XVIII, University of Missouri, Columbia, p 514
255. Zimmermann H-P, Röderer G, Doenges KH (1984) J. Submicrosc. Cytol. *16*: 203
256. Zimmermann H-P, Doenges KH, Röderer G (1985) Exp. Cell Res. *156*: 140
257. Faulstich H, Stournaras C, Doenges KH, Zimmermann H-P (1984) FEBS Lett. *174*: 128
258. Röderer G (1985) Biologie in unserer Zeit *15*: 129
259. Maddock BG, Taylor D (1980) The acute toxicity and bioaccumulation of some lead alkyl compounds in marine animals. In: Branica M, Konrad Z (eds) Lead in the marine environment, Pergamon, Oxford, p 233
260. Walsh TJ, Tilson HA (1984) NeuroToxicol. *5*: 67
261. Konat G (1984) NeuroToxicol. *5*: 87

Aluminium

Alfred Steinegger[1], *Urs Rickenbacher*[2] and *Christian Schlatter*[3]

[1] Swiss Aluminium Ltd., Feldeggstr. 4, CH-8034 Zürich
[2] Sandoz Ltd., P.O. Box, CH-4002 Basel
[3] Institute of Toxicology, Swiss Federal Institute of Technology and University of Zurich, Schorenstr. 16, CH-8603 Schwerzenbach

Summary

Aluminium, which is being produced for 100 years now, has only recently attracted the special interest of environmental chemistry. Due to changes in the natural environment, in particular due to acid rainfall, toxic effects in water organisms have become apparent that are due to greater amounts of the aluminium being set free from the large quantities in the earth's crust. Research into the conditions under which this separation of aluminium takes place, the aluminium compounds in the water and also the occurrence and mechanisms of the biological effects is only just beginning. Ecology and ecotoxicology of aluminium are, therefore, of great interest today.

Aluminium is also used in human medicine in particular on patients with chronic kidney insufficiency. They have to ingest large quantities of aluminium hydroxides in connection with dialysis treatment to bind phosphate. Occasionally disorders in the central nervous system develop in such patients, the so-called dialysis encephalopathy. Whether aluminium compounds also play a role in the case of other neurological disorders, is still very questionable. In the light of this knowledge, occupational exposure to aluminium compounds is becoming increasingly important. In view of the very low absorption of aluminium compounds in the body, there have been hardly any reports of toxic damage due to aluminium at the workplace.

Discovery, Production and Use of Aluminium and Its Compounds

Under the Latin name Alumen, which means a substance with an astringent taste, an aluminium-containing mineral has been used, by among others the Romans, Greeks and Egyptians, for 4000 years as a mordant for dyes, as an agent for the purification of water, as a substance for treatment of hides and, dissolved in water in combination with honey, to heal ulcers. However, it was only recognized in the 18th century that this mineral is a double salt consisting of aluminium and mostly potassium in the form $Me^{+1} Al(SO_4)_2 \cdot 12 H_2O$ [1]. Although aluminium is the third most frequent element in the earth's crust and is present in conjugation with oxygen and silicon in almost 250 minerals, metallic aluminium is non-existent in nature because of its high affinity to oxygen and was unknown as such for a long time. In 1746 J. H. Pott first succeeded in producing alumina, Al_2O_3 from alum (aluminium sulfate hydrate, $Al_2(SO_4)_3 \cdot 18 H_2O$). It was recognized that at the time it must be the oxide of a previously unknown metal. The presentation in a relatively pure form was first made by H. C. Ørsted in 1825 and F. Wöhler, the latter determining a few of the properties of this metal for the first time in 1845. Wöhler produced aluminium by adding anhydrous $AlCl_3$ to molten potassium. In France the cheaper sodium was substituted for potassium, whereupon it became possible to produce metallic aluminium in larger quantities. Nevertheless the production costs were still enormous — 1 kilogram of aluminium cost approx. 3000 gold francs — so that Napoleon III had dinner served in aluminium dishes in honour of illustrious guests. With the discovery of the dynamo in 1867, aluminium production became more common and cheaper: in 1886 P. L. T. Héroult in France and C. M. Hall in the USA had a process patented independent of one another, in which aluminium is produced by electrolysis of a molten solution of alumina in cryolite (Na_3AlF_6).

The first aluminium plants working according to the Hall-Héroult process were built in Pittsburgh (USA) and in Neuhausen (Switzerland) in 1888. In 1889 the annual world production of aluminium amounted to 71 metric tons [2, 3] and in 1988 to about 18 000 000 metric tons [4,5].

Bauxite, the main source of aluminium, consists of a mixture of different minerals which form due to the weathering of aluminium-containing rock. Aluminium is present in bauxite either as trihydrate gibbsite ($Al_2O_3 \cdot 3 H_2O$) or as monohydrate boehmite and diaspore ($Al_2O_3 \cdot H_2O$). Alumina is extracted from the bauxite in accordance with the so-called Bayer process. The residues are called red mud — mainly a mixture of NaAl silicate, Fe_2O_3 and TiO_2. Aluminium is a metal with low density, high reflectance and an electrical conductivity (at a purity of 99.99 %) of more than 64 % compared with copper. The thermal conductivity is 0.52 to 0.56 cgs units, the melting point of the pure metal is 660 °C. In spite of its electronegative potential of -1.67 V and the high affinity to oxygen, aluminium is very corrosion-resistant, as it forms a fine oxide coat on the surface which is insoluble in water and prevents further oxidation. Aluminium is of great economic importance. The main fields of application for metallic aluminium are: transport — it is increasingly being used for aircraft, also for railcars, automobile and truck construction —, furthermore for packaging and in the building industry, and to some extent aluminium is combined with other materials. It is also used in electrical engineering and electronics and partly also as a substitute for copper. Finely dispersed aluminium powder reacts violently with oxygen in the air.

In general aluminium has a valence of 3^+. Aluminium does occur in some compounds with the valence of 1^+, but these compounds only exist in gaseous form at very high temperatures. Since aluminium is amphoteric, with acids it forms salts such as for example chloride, nitrate and sulfate and with strong bases aluminates ($NaAlO_2$, $KAlO_2$).

Aluminium oxide is produced in various crystalline forms. While before above all the α-modification was used for aluminium production, today more and more the so-called mixed aluminas with different crystal structures are used. Most of it, approx. 33 million tons per year, is used for primary aluminium production [6, 7]. More than 2 million tons per year [6] are used as special alumina for corundum production (hardness 9 on Mohs scale), the ceramic and refractory industry, oxide and fine ceramics, for the polishing medium industry, for granulated and powdery active alumina, as white pigment for paper and cardboard, for the paint and lacquer industry, matting agents, catalyzers, catalyzer bases, molecular screens and as a filler for toothpaste. Aluminium hydroxide is of medical significance since it is taken orally sometimes in high doses as an antacid and as a phosphate binder. An important compound is aluminium sulfate $Al_2(SO_4)_3 \cdot 9\ H_2O$ or $Al_2(SO_4)_3 \cdot 18\ H_2O$; it serves as a coagulant in water purification and is also used as stabiliser in foodstuffs, for glueing paper and as a basic product for other aluminium compounds. Aluminium chloride anhydride, $AlCl_3 \cdot 6\ H_2O$, is used as a catalyst in organic chemistry and for refining petroleum. $AlCl_3 \cdot 6\ H_2O$ is used above all in the cosmetic industry as an astringent. Double salts such as sodium aluminium phosphate and potassium aluminium sulfate are much used food additives. Organic aluminium compounds are very unstable and can ignite spontaneously in contact with air and water. Alkyl aluminium is used as catalyst in the production of low pressure polyethylene [8].

Aluminium compounds were thought to be biologically inert for a long time. More recent tests have shown, however, that biologically active aluminium species may have an ecotoxicological significance. Furthermore aluminium compounds proved to be neurotoxic in animals.

Aquatic Aluminium

Chemistry of Aqueous Aluminium*

Aluminium^{+3} hydrolyzes as amphoteric ion in acid solutions to cationic and in basic solutions to anionic hydroxide aluminium complexes [9, 10]. The aquo aluminium ion $Al(H_2O)_6^{3+}$ has, according to Hem and Roberson [9], an octahedral structure and is the most frequent species of aluminium in acidic solutions. With increasing pH, further monomeric complexes arise by hydrolysis, which, however, show a marked tendency to polymerization through the formation of double OH^- bridges between the aluminium ions. The hydrolysis of the monomeric complexes takes place as follows [10, 11]:

Equilibrium constants

$$Al(H_2O)_6^{3+} \rightleftharpoons Al(OH)\,(H_2O)_5^{2+} + H_3O^+ \qquad 1.10^{-5}$$

$$Al(OH)\,(H_2O)_5^{2+} + H_2O \rightleftharpoons Al(OH)_2\,(H_2O)_4^+ + H_3O^+ \qquad 1.7\ 10^{-5}$$

$$Al(OH)_2\,(H_2O)_4^+ + H_2O \rightleftharpoons Al(OH)_3\,(H_2O)_3^0 + H_3O^+ \qquad 0.35\ 10^2$$

$$Al(OH)_3\,(H_2O)_3^0 + H_2O \rightleftharpoons Al(OH)_4\,(H_2O)_2^- + H_3O^+ \qquad 1.11\ 10^{-14}$$

The formation of the hydrogen ions leads to an acid reaction of the solution. From the balance reaction of the first hydrolysis step it results that a solution of 10^{-5} M aluminium approximately shows the pH value 5. The rest of the hydrolysis reactions is shifted to the right if sufficient hydroxy ions are present which react with the hydrogen ions. Therefore, the continuous curve of the hydrolysis is correlated with high pH value of the solution. The monomeric model of Marion et al. [12] says that soluble aluminium is represented by various monomeric aluminium species, the distribution of which is a function of the pH value (Fig. 1). With increasingly higher hydroxyl concentration, however, there is a greater tendency to form large polymers, which, arranged in parallel sheets, precipitate as gibbsite.

Hem and Roberson [9] were able to show that gibbsite particles of 0.1 μm form in 10 days at 25 °C. Until complete crystallization, however, a year or more can pass. Ageing of a buffered solution with aluminium leads to a continuous decline in monomeric aluminium species, however, and to an increase in the polymeric species [11].

Because of the amphoteric properties of the aluminium ion, the solubility of amorphous $Al(OH)_3$ and gibbsite is a function of the pH value: a decrease of one pH-unit provokes a 1000 times increased solubility below a pH of 5.6, at a pH value higher than 5.6 the solubility increase is about 100 times per one pH-unit [13].

In natural waters, the behaviour of aluminium, however, is more complicated, because complexes of partially remarkable stability are formed not only with hydroxyl ions but also with different other inorganic and organic ligands [11, 14] (Table 1).

---

* The term aluminium stands for all kinds of aluminium compounds.

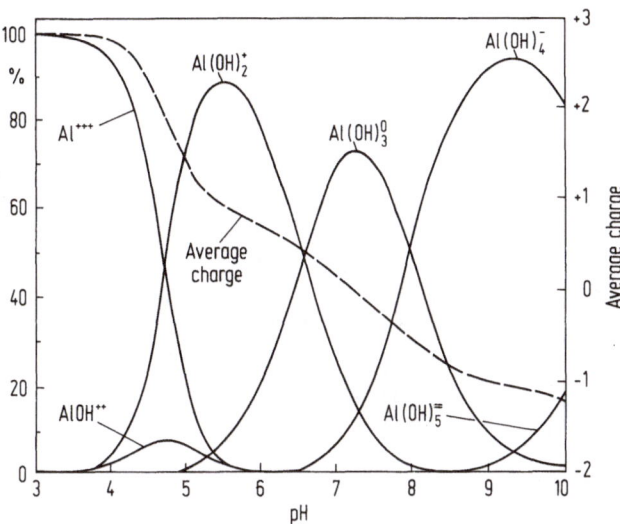

Fig. 1 Relative distribution and average charge of the total soluble aluminium species as a function of pH, ionic strength = 0.1 M [12].
Reprinted by permission of G. M. Marion, D. M. Hendriks, G. R. Dutt and W. H. Fuller, "Aluminium and silica solubility in soils", Soil Science, vol. 123, p. 82, (C) by Williams & Wilkins, 1976

Table 1. Equilibrium constants of complexes with aluminium and different inorganic and organic ligands [11, 14]

	Equilibrium constant
$Al^{3+} + F^- = AlF^{2+}$	1.05×10^7
$AlF^{2+} + F^- = AlF_2^+$	5.5×10^5
$AlF_2^+ + F^- = AlF_3$	1.86×10^4
$AlF_3 + F^- = AlF_4^-$	5.0×10^2
$AlF_4^- + F^- = AlF_5^{2-}$	15.5
$AlF_5^{2-} + F^- = AlF_6^{3-}$	0.9
$Al^{3+} + SO_4^{2-} = AlSO_4^+$	1.6×10^3
$AlSO_4^+ + SO_4^{2-} = Al(SO_4)_2^-$	80
$2Al^{3+} + 2H_4SiO_4 + H_2O = Al_2Si_2O_5(OH)_4 + 6H^+$	5.25×10^{-12}
$Al^{3+} + C_2O_4^{2-} = AlC_2O_4^+$	1.0×10^7
$Al^{3+} + C_7H_4O_3^{2-} = AlC_7H_4O_3^+$	1.0×10^{14}

Since some complexes have a high stability, it can be assumed that aluminium is largely present sequestered in natural waters [9]. Due to the sequestering, the solubility of $Al(OH)_3$ is increased. The concentration of organic acids in eutrophic waters can further increase the solubility of aluminium compounds [15] (Fig. 2).

Aluminium in Natural Water

In near-neutral waters, the content of dissolved aluminium is generally low i.e. less than 100 µg/L. (Table 2[16, 17]). The concentrations of dissolved aluminium are mostly

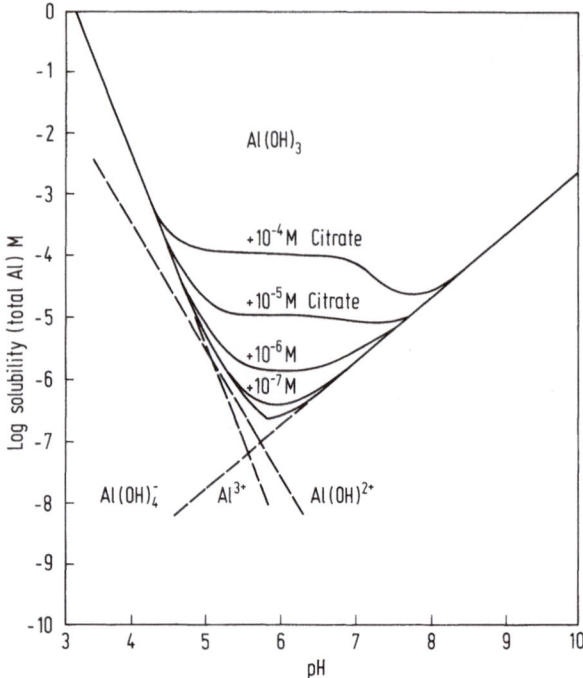

Fig. 2 Theoretical solubility of $Al(OH)_3$ in the presence and absence of citrate [15].
Reprinted by permission of Y. R. Chen, J. N. Butler and W. Stumm, "Adsorption of phosphate on alumina and kaolinite from dilute aqueous solutions", Journal of Colloid and Interface Science, vol. 43, no. 2, p. 434, (C) by Academic Press, 1973

lower in seawater than in freshwater, since the solubility of aluminium decreases with increasing salt content. A number of the published measurements of aluminium in waters refers to the total aluminium content with or without suspended fraction [3, 11]. To obtain biologically reliable results, measuring methods are required, however, which allow differentiation into monomeric, polymeric and inorganic or organic

Table 2. Aluminium concentrations in nearly neutral waters [16, 17]

Locality	Dissolved Al ($\mu g/L$)	Al-Species	Method	Water Parameter	Ref.
Clear Lake USA	1.1	monomere	spectrophotom. of Al-8-hydroxy-quinoline	pH = 7.0	[16]
Sherwood L. USA	44.9	monomere	ditto	pH = 6 eutrophic	[16]
Yellow River USA	26	monomere	ditto	pH = 7.7	[16]
Conway River UK	14.3	filtered (0.45 μm)	fluorimetry of Al-lumogallion	pH = 7.1	[17]
North Sea	1.5 (mean)	filtered (0.45 μm)	ditto	salinity >34%	[17]

complexed aluminium, and a sample pretreatment, which does not destroy the original speciation in the sample. Complexing of aluminium leads in natural waters to higher concentrations than would be expected because of the solubility of gibbsite. Mainly organic acids serve as ligands. The connection between high contents of organics such as humics and high aluminium concentrations has been shown a number of times [18]. An examination by Budd et al. [19] shows a significant correlation between acid reactive aluminium from different rivers and dissolved or colloidal organic content. Johnson et al. [20] stated that most organically complexed aluminium is associated with 10^3–10^4 mol. wt. fraction of dissolved organic carbon.

The most influential water quality criterion which can contribute to the mobilization of aluminium in waters, is, however, the pH value.

Aluminium in Acidic Waters

In waters with a pH 5–6, higher aluminium contents are frequently observed. Examples for natural acid waters are acid spring waters and marshes [18]. Acidification can for example arise due to oxidation reactions [21]. Acidic inputs which can lead to an acidification of the water, are industrial wastes, mining and acid precipitation. In Scandinavia, Scotland, South-East Canada and the North-East of the United States, the increasing acidification of freshwater is mainly due to acid precipitations with volume weighted mean annual H^+ concentrations of at least 25 µeq/L [22]. Due to the resulting drop in pH, particulate aluminium is dissolved to a greater degree. Most of the aluminium in acidic waters probably stems from the percolate of the soil zone [20, 23].

Johnson et al. [20] described the effects of acid precipitations as a two step chemical reaction: first of all the hydrogen-ionic acidity is neutralized by the dissolution of aluminium. The aluminium buffering leads to an aluminium acidity stage with a pH value of 4.7–5.2. In the second step, the hydrogen acidity and the aluminium acidity are neutralized by chemical weathering products such as calcium and sodium. Since the chemical weathering reactions take place much more slowly than the dissolution of aluminium, the aluminium acidity stage is a long term and possibly global phenomenon ([24] Fig. 3).

Driscoll et al. [24] examined the seasonal variations of aluminium contents in stream waters differentiating between the following three fractions of the total soluble aluminium: the labile, monomeric aluminium which, besides the aquo aluminium, comprised further inorganic complexes such as aluminium fluoride, hydroxide and sulfate, the non-labile monomeric aluminium (organically complexed) aluminium and finally the total (acid soluble) aluminium. The fractions reacted differently to seasonal changes of pH levels and the total organic carbon concentration (TOC) (Fig. 3).

Inorganic monomeric aluminium followed the pH value inversely, the latter varied mainly because of hydrological events (rainfall and melting snow) or biological processes. Organic aluminium seemed to be independent of the pH value but correlated with the TOC.

Although acidification of the water mainly occurred in areas with fewer buffering minerals in the bedrock, Bjärnborg [25] observed that hydrological events such as for example melting snow can also lead to reductions in the pH value for a short time and

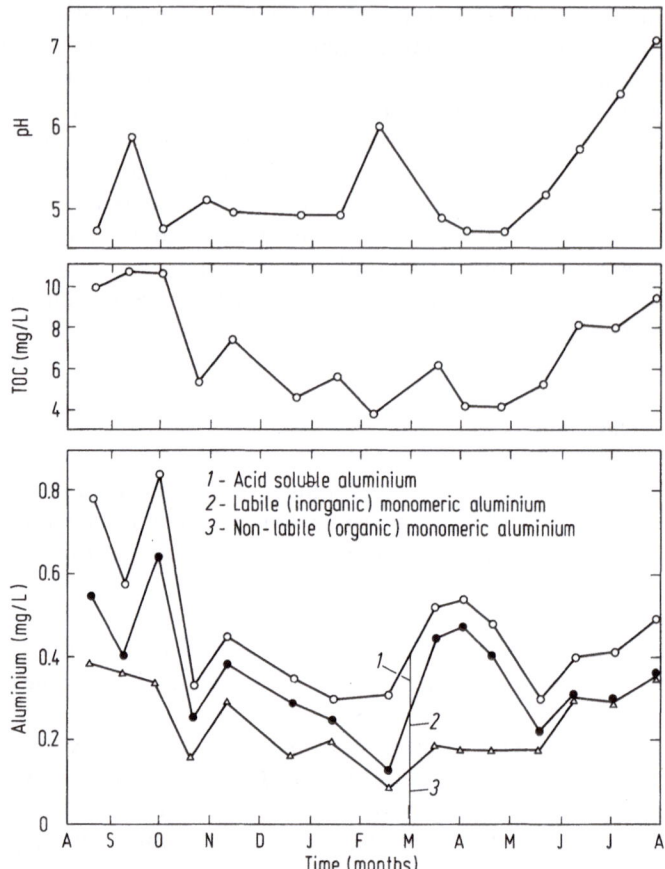

Fig. 3 Temporal changes (August 1977–August 1978) in pH: total organic carbon (TOC) and aluminium forms in a tributary of a lake: low pH and high levels of labile monomeric (inorganic) aluminium after rainfalls and during snow-melt. Under low waterflow conditions and elevated pH values decreased labile monomeric (inorganic) aluminium levels are observed; TOC and non-labile inorganic aluminium high in autumn and summer [24].
Reprinted by permission of C. T. Driscoll, J. P. Baker, J. J. Bisogni and C. L. Schofield, "Effect of aluminium speciation on fish in dilute acidified waters", Nature, vol. 284, p. 162, (C) by Macmillan Magazines Ltd., 1980

to a drastic increase in the soluble aluminium in surface waters in areas with calcerous soils and buffered waters.

Terrestrial Aluminium

Aluminium is the third most frequent element in the earth's crust after oxygen and silicon.

By far the largest amount of aluminium is present in the stones in the earth's crust. Basalt for example contains an average 87 600 mg Al/kg and granite 77 000 mg/kg [26].

When the rocks are weathered, soluble cations such as calcium, potassium, magnesium and sodium are largely washed out while aluminium accumulates in the soil in the form of aluminium silicates and hydratized oxides [10]. This leads to very high total aluminium quantities in the soil: the values are between 5000 and 60000 mg/kg. In extreme cases up to 300000 mg/kg are reached [3]. Aluminium is largely immobile in the soil. Aluminium silicates and to a limited extent also hydrated aluminium oxides have an important function as cation exchanges in the soil [10].

The content of mobile aluminium is compared to the total aluminium content in the soil, very small: in a soil with the pH value of 5.6, Webber et al. [27] measured 800 mg/kg of extractable[1] or 6 mg/kg exchangeable[2] and 0.6 mg/kg dissolved[3] aluminium. In chalky soils with pH > 6, measurements by MacLean et al. [28] showed only 0.8 mg/kg exchangeable and 0.09 mg/kg soluble aluminium.

Mobilization of Aluminium in the Soil

The exchangeable aluminium is mainly made up of monomeric aluminium species, because the polymeric aluminium species adhere too strongly to the surface of the soil colloids. In addition aluminium precipitates as gibbsite over the entire pH area as soon as the solubility product of this mineral is exceeded. The soil pH must reach a value of 6 before significant quantities of exchangeable or soluble aluminium can be expected [10]. Besides the pH in the soil, other factors influence the mobilization of aluminium: Wada and Wada [29] showed that reactions between hydroxy aluminium and orthosilicic acid can lead to the formation of hydroxy aluminium silicate soluble in water. Ligands such as organic acids (malate, oxalate, citrin etc.) and organic polymers (fulvic acids and humic acids) also mobilize aluminium due to the increased solubility and transport it into deeper layers in the soil [10, 12, 13].

Contrary to monomeric aquo aluminium, organo aluminium complexes should not be exchangeable but only extractable [30]. The differentiation is made between exchangeable and non-exchangeable aluminium on the basis of the applied methodology or the extraction agent used [27, 28, 30—33].

Aluminium in Acid Soils

Acidification of the soil is a global problem with different causes. The mechanisms which lead to an acidification of the soil are mostly internal soil events e.g. hydrolysis of soil minerals, the oxidation of sulfides such as pyrite (FeS_2), the use of ammonium and phosphate-containing fertilizers, decomposing of plant residues and organic wastes into organic acids [10]. CO_2 from the decomposition and from the root respiration combines with water to H_2CO_3 and is also a source for H^+ in a lot of forest soils [34]. An external source is the atmospheric hydrogen input (acid precipitation). However, compared to internal H^+ formation, the atmospheric contribution seems to be relatively small. It was estimated at approx. 10% for badly exposed relatively small areas [34]. While in neutral soils, calcium, magnesium, potassium and so-

---

[1] Aluminium, which can be washed out with a complexing and/or acid solution.
[2] Aluminium, which can be eluted with a neutral salt solution.
[3] Aluminium soluble in 0.02 M $CaCl_2$ or in water saturated soil.

dium are the most frequent cations, in acid soils aluminium and manganese are predominant cations [31].

The acidification of the soil with a drop in the pH value by one unit and more and the resultant mobility of aluminium has been found to be due to acid precipitation on various occasions [30]. This correlation only applies in those extraordinary cases, however, in which the atmospheric H^+ entry reaches or exceeds the soil internal H^+ production [22, 31]. Ulrich et al. [31] described a combined input by acid precipitation (deposition rate between 1.1 and 2 kmole $H^+ \cdot ha^{-1} \cdot yr^{-1}$) and an internal H^+ formation (production rate between 2.9 and 5.5 kmole $H^+ \cdot ha^{-1} \cdot yr^{-1}$) in the badly polluted Solling, West Germany. From that a soil pH (measured in $CaCl_2$ solution) results of between 3 and 4, two thirds of the H^+-buffering having taken place through the dissolution of polymeric hydroxy aluminium. At a depth of 20–50 cm, the aluminium content in the soil solution rose from 10% of the equivalent cation sum in 1966 to 20 to 30% in 1979.

Analytical Methods

Various methods have been used for the analysis of aluminium: chemical methods, emission spectrometry, atomic absorption spectrometry with flames or graphite tube techniques, neutron activation, X-ray fluorescence and fluorimetry.

Atomic absorption spectrometry is the most frequently used method. The advantages of this method are: it is quick, has a high specificity, few matrix interferences, relatively high sensitivity. In spite of the advantages of the method, the published values diverge e.g. normal plasma values by the power of ten (see: Normal plasma aluminium values, Table 6). Erroneous measurements occur in most cases due to contaminations. This is a critical point in aluminium analysis, since aluminium occurs in measurable concentrations virtually everywhere.

There are innumerable possible sources of contamination. The most important aluminium containing materials which can come into direct contact with the samples, e.g. glassware, low pressure polyethylene, and filter papers can contain aluminium in the ppm (µg/g) range. Relatively aluminium free materials such as high pressure polyethylene, polypropylene, teflon and quartz should be used, therefore. Laboratory materials made of these materials can be contaminated on the surface and should be cleaned before use, for example with acidified bidestilled water or saturated EDTA solution.

Toxic Effects on Plants

In acid soils, aluminium is the limiting factor for the growth of crops. Foy et al. [35] report that the damage from aluminium is often difficult to recognize: the changes in the leaves are similar to the symptoms in the case of phosphorus or calcium deficiency. The most marked morphological changes take place, however, at the roots, whose function is badly affected. The physiological mechanisms of aluminium toxicity vary, e.g. arresting the transportation of various elements such as calcium and phosphorus, restriction of the cell divisions in secondary roots, arresting the respiration of roots, effects on cell walls and complexation of nucleic acids.

Plant species and varieties are sensitive to soluble aluminium in the soil to different degrees: the root growth of several vegetable plants was adversely affected as from 0.5 mg Al/kg in the soil solution [36]. Various varieties of barley proved in an experiment by MacLeod and Jackson [36] to be sensitive to aluminium concentrations between 0.75 and 2.2 mg Al/kg. However, species more tolerant to aluminium are known and also species whose growth are even furthered by certain quantities of aluminium. The sensitivity is often due to genetic reasons and as in the case of wheat for example is dependent on a few genes and is based on the ability of the plant to complex and precipitate aluminium [33].

Liming has proved advantageous to reduce the quantity of exchangeable aluminium and, therefore, the toxic effects. According to MacLeod and Jackson [37] the subsoil is only slowly influenced by liming and the arrest of root growth can continue in this range.

While there exists an acceptable data base of the effects of aluminium on crops, examinations regarding forest trees are rarer and more difficult to interpret. In a field study hardly any toxic effects were found on the Norway spruce, the Scots pine and birch in the case of an artificial acid input which corresponded to 10 to 30 times the annual input in the polluted area in Solling, West Germany and with a content of soluble aluminium of up to 60 mg/L soil solution [34]. Examinations of other authors [38], however, showed necrosis in young beech and pine roots induced by aluminium. Tischner et al. [38] investigated the effect of aluminium in the nutrient solution of pine germs. In the case of an aluminium concentration of 27 mg/L and a pH of 4, symptoms on the roots occurred to a greater degree and a drop in the chlorophyll content and the chlorophyll a/b ratio was observed. Furthermore a smaller fresh weight was ascertained. The authors discuss these results in connection with acid precipitations and local damage to forest trees.

Toxic Effects on Aquatic Biota

In acidic waters, a reduced number of animal species and occasionally total loss of fish stocks are observed [39, 40]. Mobilized aluminium is one of the most important reasons for harmful effects to aquatic biota in acidic waters [11, 23, 41, 42].

Schofield and Trojnar [43] investigated the survival of brook trout stocked into 53 acidic Adirondack lakes as a function of 12 water quality parameters. Covariate analysis indicated that the concentration of total aluminium was the primary chemical factor controlling fish stocking success. High concentrations of soluble aluminium in effluent discharge downstream from plants or water treatment works, which used aluminium as a coagulant, may have caused fish to die [44]. The target organ of aluminium toxicity to fish seems to be the gill tissue where ion exchange is impaired resulting in a loss of plasma chloride together with respiratory distress caused by mucus clogging of the gills and hyperplasia of the gill tissue [45]. Effects of hydrogen ion stresses are very similar but the aluminium toxicity becomes visible at a pH value which is more than one unit above the value normally harmful for fish [39, 46]. Furthermore Baker and Schofield [41] showed that the sensitivity to low pH levels increases with increasing age while the sensitivity to elevated aluminium levels decreases with increasing age.

This observation is contradicted by Brown [46] who found a higher sensitivity to aluminium with 1 year old than with 1 month old trout.

Muniz and Leivestad [39] studied the loss of plasma chloride in brown trout at different pH levels with and without added aluminium. Without added aluminium, no response was found at pH 4.6 and higher. Exposed to 0.9 mg Al/L effects occurred at pH 4.6 with a maximal response at pH 5.1–5.5. The fact that the toxic effects of aluminium in the equilibrium reaction stage are most marked at pH 5, was confirmed by various other tests [24, 33].

The dissolved aluminium species are of different toxicity: organic complexation and to a smaller extent, inorganic complexation mitigate aluminium toxicity [24, 47]. Baker [47] found, contradictory to the results above, higher toxicity of aluminium in natural waters than in laboratory conditions although they had almost identical aluminium concentrations and pH values. This could imply that organic aluminium compounds could be toxic nevertheless. The most toxic fraction seems to be the free, inorganic aluminium. The knowledge of the content of total dissolved aluminium only gives insufficient information about the potential toxicity. Table 3 shows the findings of some experimental toxicity studies with aquatic organisms [11, 24, 46, 48, 49].

Table 3. Toxic effects of aluminium on aquatic organisms

Animal	Water	pH	Total Al µg/l	Al-Speciation	Effects	Ref.
Brook trout	synthetic	5.20	20	free, monomeric	mean survival 99%	[24]
Salvelinus	soft water	5.20	420	ditto	mean survival 28%	
fontinalis		4.40	480	ditto	mean survival 42%	
		5.20	500	aqueous inorg. complex (fluorid)	mean survival 45%	
		5.20	500	organic complex (citrate)	mean survival 96%	
Brown trout	tap					
yolk sac fry	0.5 mg Ca/L	5.4	500		survival about 30%	[46]
Salmo trutta	2 mg Ca/L	5.4	500		survival 0%	
	2 mg Ca/L	5.4	250		survival over 90%	
	0.5 mg Ca/L	5.4	250		survival about 20%	
White sucker	softened					
Catostomus commersoni						
Embryo	tap	5.2	160	as monomeric 140	hatching 0%	[41]
Larvae		5.4	40	as monomeric 50	survival 0%	
Postlarvae		5.0	90	as monomeric 80	survival 11%	
Postlarvae		5.4	70	as monomeric 60	survival 80%	
Goldfish	hard	5.5	10 500		lethal 1 to 10 h	[11]
Carassius auratus auratus		6.8	10 500		lethal 12 to 99 h	
		7.6	1 000		not lethal in 2 h	
Polychaete	filtered seawater	7.6 to 8.0				
Capitella capitata			2000		96 h LC_{50}	[48]
Ctenotrilus serratus			480		96 h LC_{50}	
Crustacea						
Daphnia magna aged		6.5 to	1 400		3-week LC_{50}	[49]
	freshwater	7.5	680		50% reprod. impair.	
			320		16% reprod. impair	

Liming reduces the aluminium concentration in water [47]. On the other hand aluminium has been reported as a possible intoxicant in cases where acidified lakes had been limed in order to increase the pH. The fish deaths observed were probably a result of the enhanced desorption of aluminium from the sediment in the form of basic aluminium complexes [50].

Brown [46] and Dietrich et al. [51] showed that calcium and NaCl, respectively counteract the toxic effect of liming on the gills. Eriksson et al. [40] state that liming has positive effects on fish stocking provided that the pH level can be kept stable. However, that is hardly possible in regions with spring tides.

Besides the toxic effects on aquatic biota, there are biochemical effects in connection with elevated aluminium concentrations in acidified lakes: labile inorganic aluminium precipitates phosphate with the result that against expectations acid lakes prove to be low in phosphate. In addition aluminium inhibits the hydrolysis of phosphorate esters and, therefore, reduces the availability of phosphate for planktonic organisms which in turn react with an increase of phosphatase production. That is why a higher phosphatase activity is measured in acid, oligotrophic lakes. In eutrophic (humic) lakes, the percentage of inorganic aluminium, is too small, however, to reduce the phosphate concentration [52].

Kinetics of Aluminium in Higher Organisms

Intake of Aluminium

Most of the aluminium absorbed orally by man stems from the natural content of vegetable foodstuffs, which can amount up to 0.2 g/kg dry weight [53]. The contents can also be higher (> 1 g/kg) in plants which accumulate aluminium such as for example certain spices, tea, and asparagus [54] or in plants which grow on acidic soil and store exchangeable aluminium to a level almost independent of the corresponding concentration in soil. Attanandana et al. [55] report that rice plants in Thailand which grow in moderate acid soils with a content of 0.8–7.3 mg/100 g aluminium exchangeable with 1 N KCl solution, contained 0.21–0.23 g of aluminium/dry weight while rice plants from acid soils with a content of 103 mg exchangeable aluminium/100 g had an aluminium content of 0.3 g/kg. On the subject of the contamination of foodstuffs with aluminium when cooking or from storage in aluminium utensils, numerous studies, mainly older studies have been reviewed by Campbell et al. [1]. Generally hardly any aluminium is absorbed by neutral foodstuffs. In acid or alkaline foodstuffs, the cooking process in aluminium utensils can raise the aluminium content up to a factor of 20. The quantities absorbed are of little importance when compared with the natural contents in specific vegetable foodstuffs. They only rarely exceed the range of concentration of the raw products. Furthermore the mobilization of aluminium is considerably reduced by adding sugar [1].

The aluminium content in foodstuffs of animal origin is much smaller at 1–2 mg/kg dry weight than in vegetable foodstuffs. The contents in beverages such as wine (6.8 mg/L), beer (0.5 mg/L) and milk (0.5 mg/L) are small in comparison to vegetables [3, 54].

Measurements of the aluminium concentrations in the drinking water of 50 towns in the USA showed values between 3.3 and 1500 µg/L. Contents in the drinking water of more than 100 µg/L mainly arise due to the use of aluminium sulfate as coagulant in drinking water preparation [56]. Also such values do not contribute unduly to normal daily oral intake of aluminium, they can become harmful when used for home dialysis. Another manmade source of aluminium is the use of different aluminium compounds as baking powder, anticaking agent, pickling salt etc. [57]. The contribution which the aluminium-containing food additives make to the total daily aluminium intake is probably very small: if aluminium containing baking powder is used for example, about 5 mg of sodium aluminium sulfate per portion of cake results [57]. That corresponds to an intake of 0.4 mg aluminium. The tolerable daily intake of sodium aluminium phosphate, which is also used in baking powder, was fixed by WHO at max. 6 mg/kg body weight [56, 58].

Estimates of the daily aluminium intake via foodstuffs are 5–135 mg. An estimate of the average intake was given with 20 mg [56, 57]. In a recent total diet study from Switzerland, a daily intake of 1–7 mg with an average of 3.4 mg was determined [59]. Similar results were obtained in a German study [53]. Among individual food items, tea probably represents the most important single source of aluminium [60]. The aluminium content is usually in the order of 2–6 mg/L [60]. The recently published result of Coriat and Gillard [61] showing typical contents of 40–100 mg/L is difficult to explain.

Aluminium hydroxide is widely used as antacidum against hyperacidification of the stomach (antacida) and gastric ulcers. Uremic dialysis patients take aluminium hydroxide as phosphate binder in quantities up to 2 g per day. Some salicylic acid-containing analgetica contain aluminium hydroxide as buffer substance. The normal daily dose leads to an aluminium intake of 10 to 100 mg. In the case of a high dose as is customary for example for arthritis patients, the intake can amount to approx. 800 mg of aluminium [51]. The aluminium intake can be factors higher due to its application medically than through foodstuffs. However, it must be taken into account that in the medical sector mainly aluminium hydroxide is used and this compound, which does not easily dissolve, is only absorbed gastrointestinally in very small quantities (see next chapter).

Various aluminium compounds are used in medicine: vaccines against tetanus, diphtheria etc. contain aluminium compounds (mainly aluminium hydroxide) as adsorbant in order to slow down the release of antigenes. The activity of vaccines is increased as a result. The US Food and Drug Administration (FDA) limited the amount to 1.14 mg aluminium/dose.

Absorption, Excretion and Retention

Aluminium compounds which dissolve easily, are absorbed better than those which are only sparingly soluble. Due to the solubility properties of its compounds, aluminium is mainly absorbed in the acid environment of the stomach and the proximal duodenum [62]. The gastrointestinal absorption is influenced by various factors: Mayor et al. [63] observed that parathyroid hormone furthers the gastrointestinal absorption of aluminium. Feinroth et al. [64] showed that the intestinal absorption was blocked in an in vitro system by dinitrophenol, glucose deficiency and low temperatures. The authors concluded that the absorption is energy dependent and requires an active carrier. While examinations of the gastrointestinal absorption of aluminium

Table 4. Gastrointestinal absorption and renal excretion of various aluminium compounds on healthy subjects [62, 65].

Al compound	Total Al Ingested (g)	Days of Treatment	Number of Subjects	Plasma Al before Treatment µg/L	Plasma Al after Treatment µg/L	Urinary Loss before Treatment µg/day	Urinary Loss after Treatment µg/day
Al hydroxide [65]	2.2	3	13	6 ± 3	17 ± 3	16 ± 8	275 ± 174
Al phosphate [65]	2.2	3	6	7 ± 4	9 ± 2	15 ± 6	60 ± 35
Al lactate [62]	2.5	20	8	4 ± 1	7 ± 1	36 ± 5	117 ± 19

hydroxide were carried out frequently on dialysis patients, studies with healthy people using a reliable analysis technique, are rare. The only slight rise in the plasma aluminium concentration during the treatment of healthy people with aluminium compounds shows the efficiency of the gastrointestinal barrier, which only lets small quantities of aluminium through ([62, 65], Table 4).

Aluminium binds phosphate in the basic environment of the intestine and is discharged with the stool. The excretion of absorbed aluminium takes place through the kidneys. The mechanism of the discharge was visualized by microscopy [66]: aluminium is absorbed by the lysosomes in the proximal kidney tubuli and precipitates there as aluminium phosphate. This insoluble salt is then discharged into the tubular lumen and is carried away with the urine.

After the oral administration of 125 mg aluminium as a lactate, no retention was found [64]. Gorsky et al. [67] ascertained a retention in the range of 23 to 313 mg aluminium per day when they administered very much higher doses. This investigation was, however, carried out with people who had partially osteoporosis and kidney damage. A changed metabolism of the electrolytes and also of aluminium in this group is not excluded, therefore. In experiments with laboratory animals, accumulation in organs was shown on various occasions: in an experiment of Mayor et al. [63] the aluminium concentration in the brain, bones and kidneys of rats increased after administering parathyroid hormone and aluminium simultaneously. Marquis [68] observed that 0.06 and 0.12% aluminium in drinking water increased the aluminium concentrations in the brains of adult rats from 3.1 µg/g to 5.1–6.7 µg/g. The offspring of the animals treated did not have any increased contents in the brain. Ellis et al. [69] administered intraperitoneal (ip) doses of altogether 33 to 38 mg aluminium per rat over 52 days. The aluminium content of the whole bone (femora) from an average 15.4 ± 4.7 µg/g in the reference animals to an average of 109.3 µg/g (range 90–124 µg/g). Ip administrations of altogether 38 to 109 mg of aluminium over 45 to 85 days showed average concentrations in the bones of 163 ± 9 µg/g. Bone aluminium remained elevated up to 49 days when there were no more aluminium injections.

Raised aluminium concentrations have been recorded by Alfrey et al. [70] in brain, bone and other tissues of hemodialysis patients with chronic renal failure, who had been treated with aluminium hydroxide gels. Cann et al. [71] suspect that the concentrations of aluminium in the parathyroid glands are linearly related to dietary aluminium intake.

Harwerth et al. [72] investigated 51 workers exposed to metallic aluminium dust from 200–600 µg/m^3. The median blood level was 11.9 µg/L compared to 7.5 µg/L in a non-exposed control group of 41 persons. The corresponding urine values were 6.2 µg/g creatinine and 5 µg/g creatinine. Considerably higher levels were found by Hartung et al. [73] in 25 workers exposed to metallic aluminium powder used for pigment production. The aluminium levels in the air amounted to 4–8 mg/m^3. The median aluminium concentrations in serum were approx. 80 µg/L, that is 15 times the value of non-exposed people.

Even in situations of possible respiratory uptake such as working places in the aluminium industry, most of the aluminium is incorporated via the diet, nevertheless the lung tissue shows the highest aluminium concentrations (Table 5), the concentrations being higher in the upper lobe than in the lower lobe [83].

That indicates that from the aluminium content in the dust in the air, which amounts to 0.005–20 µg/m^3 outside industrial zones [3], certain quantities are deposited in the respiratory tract and also possibly in other tissues. In an inhalation study Stone et al. [79] observed that 0.25 to 25 mg/m^3 aluminium chlorhydrate over 24 months raised the aluminium concentrations in the lung tissue of guinea pigs or rats by less than twice with the smallest dose and 6–20 times with the highest dose compared with the reference values (30–50 µg/g).

In the case of a 49 year old worker, who worked in a ball mill room of an aluminium powder factory for 13^1/$_2$ years and who died of a terminal broncho-pneumonia following progressive encephalopathy, the lung and the brain showed 20 fold aluminium concentrations and the liver 122 fold [83].

The mean aluminium concentration in urine ranged from approx. 30–100 µg/L depending on the time of urine voiding. The concentration in urine of non-exposed persons was about 5 µg/L. Sjögren et al. [84] studied workers exposed to aluminium in four different industrial processes: electrolytic production of aluminium, production of aluminium powder, production of aluminium sulfate and aluminium welding. No difference in the blood level was found in workers in the electrolytic production of aluminium, whereas in the other groups the serum concentration was slightly higher than in the respective reference group. In a further study, Sjögren et al. [85] observed urine aluminium concentrations in the area of 200–400 µg/L in aluminium welders with exposure times of 18–23 years irrespective of immediate previous exposure. In non chronically exposed volunteers, the remaining aluminium returned to the preexposure level with an estimated half-life of about 8 hours. Hartung et al. [73] found in a group of 25 workers exposed to aluminium powder (total aluminium dust 18 mg/m^3, respirable 4 mg/m^3) a mean serum aluminium level of 80 µg/L, that is 15 times the value of non-exposed people.

Schlatter et al. [86] determined the blood aluminium levels in aluminium workers. Workers in the casthouse showed only slightly increased levels (mean 3.5 µg/L) versus 2.8 µg/L in non-exposed controls. In the aluminium electrolysis, the values ranged from 2.1 to 53.2 µg/L. The values fluctuated by a factor up to two during a working day. The levels before a shift were 5–13 times higher than those of non-exposed people.

Reports on the aluminium intake of occupationally exposed persons published before 1980 are difficult to interpret due to a lack of appropriate methodology for the determination of low aluminium levels in biological tissues. In these early days normal values were up to 2100 times higher than today. Valentin et al. [87] did not find any significantly higher aluminium content in the serum of workers in corundum producing and processing factories compared with reference persons. However, the renal aluminium elimination was twice that of normal persons. More recent results from the same author for workers exposed to aluminium powders (concentration in air 5–28 mg/m^3) showed plasma aluminium levels from 14–238 µg/L (mean 90 ± 50 µg/L). The mean aluminium plasma level of non-exposed persons was 7.2 ± 2 µg/L [88]. Similar results were obtained in aluminium welders and polishers by Mussi et al. [89]. The mean plasma levels were in the area of 5 µg/L.

Table 5. Normal aluminium concentration (µg/g or µg/ml) in different organs, tissues and body fluids. All organs stem from human beings and the aluminium concentrations refer to the dry weight. Exceptions are marked separately (a = wet weight, b = rat)

Tissue	Aluminium	Analytical Method	Ref.	Tissue	Aluminium	Analytical Method	Ref.
Liver	17[a]	NAA	[74]	Spleen	18[a]	NAA	[74]
	15.8	NAA	[75]		17.2	NAA	[75]
	4.1	AAS	[70]		2.6	AAS	[70]
	4.6	AAS	[70]		3.3	AAS	[70]
	0.7[b]	AAS	[76]				
	0.6[b]	AAS	[78]	Lung	38[a]	NAA	[74]
					43	AAS	[70]
Kidney	18[a]	NAA	[74]		67	AAS	[70]
	17.5	NAA	[75]		97.2	NAA	[75]
Whole brain	2.2	AAS	[77]		10.1[a, b]–52[a, b]	AAS	[79]
	1.9	AAS	[77]				
	1.5	AAS	[77]	Bone	3.3	AAS	[70]
	2.7[b]	AAS	[76]		5.6	AAS	[70]
	0.8[b]	AAS	[78]		10.6	NAA	[69]
					5.7	NAA	[69]
Brain white matter	0.4	AAS	[78]		1.8[b]	AAS	[76]
					15.4[b]	NAA	[69]
Brain gray matter	2.4	AAS	[70]	Cerebro-	0.035	AAS	[82]
	0.9	AAS	[78]	spinal fluid			
Heart	20[a]	NAA	[74]				
	24.1	NAA	[75]	Amniotic-	0.093	AES	[81]
	1.0	AAS	[70]	fluid			
	1.2	AAS	[70]				
	0.5[b]	AAS	[76]	Hair	1.3	AAS	[80]
Muscle	15[a]	NAA	[74]		6.2	AAS	[82]
	23.6	NAA	[75]				
	1.2	AAS	[70]				
	1.2	AAS	[70]				
	0.1–6.2	NAA	[71]				
	8	NAA	[71]				
	8[b]	NAA	[71]				
	1.6[b]	AAS	[78]				

Concentrations of Aluminium in Biological Samples

Determination of normal tissue values of aluminium in man and animal is difficult because due to the small gastrointestinal absorption of aluminium, the levels are low. On the other hand the contamination hazard during sampling and processing, because of the abundance of aluminium, is very high.

Measurements of normal values e.g. of reference plasma and different organs (Tables 5 and 6) from different laboratories differ considerably mainly for these reasons.

It can be assumed that in measurements which give more than 5 µg Al/L plasma from normal people, these external causes were not sufficiently under control. For measuring aluminium in biological material today, mostly graphite furnace atomic absorption spectroscopy (ASS) is used. This method combines an adequately high sensitivity (detection limit 2 µg/L) with a minimum of sample pretreatment, which is an important prerequisite for avoiding contamination.

Table 6. Normal plasma aluminium values (µg Al/L)

Mean	S.D.	Range	Number of Subjects	Analytical Technique	Year of Publication	Ref.
1.6	1.29		11	AAS	1982	[90]
2.1	2.2	0.0– 7.6	14	AAS	1980	[91]
<4		<2.5–10	37	AAS	1981	[90]
6.5	4.1	2 –14	28	AAS	1983	[90]
6.6	2.6	3.8–12.5	10	AAS	1984	[89]
7	2		13	AAS	1977	[62]
9.8		4.1–20	15	AAS	1981	[90]
14.2	7.1	4.0–35.4	40	AAS	1976	[87]
25		10 –50	10	NAA	1978	[92]
34.1	3.5	28 –40	20	AAS	1978	[93]
42	ˋ 16	20 –75	20	ICP	1980	[94]

AAS: Graphite furnace atomic absorption spectroscopy
NAA: Neutron activation analysis
ICP: Inductive coupled plasma atomic emission spectroscopy

Binding of Aluminium

The following observations have been made about binding of incorporated aluminium: according to Elliott [95], plasma aluminium is 60 to 70% tied to high molecular weight proteins and 10 to 20% albumin. 10 to 30% are ultrafiltrable. Trapp [96] showed that aluminium is bound to at least one of the two specific iron binding sites of serum transferrine and also to serum albumin. Galle et al. [97] found aluminium in the brain tissue of people exposed to aluminium (dialysis patients) in high concentrations in lysosomes where it was associated with phosphate. In animals exposed to aluminium, aluminium accumulates in the nuclei of brain cells, mainly in chromatine: measurements of the intranuclear aluminium content in brain cells of cats by Crapper and De Boni [77] revealed the surprisingly high value of 700 µg Al/g DNA (desoxyribo nucleic acid), aluminium being distributed on the chromatin to varying degrees: heterochromatin contains 1550 µg Al/g DNA, intermediate chromatin and euchromatin 399 and 150 µg/g DNA. In cats treated with aluminium, this concentration of the chromatin rose by 3–5 times as much. Karlik and Eichhorn [98] stated that the aluminium-DNA binding is possibly very stable since deproteinising of the chromatin only removed 12% of the chromatin bound aluminium.

Again with respect to the brain tissue, Siegel and Haug [99] were able to prove stoichiometrically binding of aluminium to bovine brain calmodulin, a multifunctional calcium regulating protein and Mazarguil et al. [100] the complex formation of aluminium with (Leu^5)-Enkephaline. Maloney et al. [101] showed that in bones of patients with renal failure and high aluminium absorptions, aluminium is found mainly in the junction of mineralized bone and osteoid.

Mammalian Toxicology

Acute and Subchronic Toxicity

Since only small quantitites of aluminium are absorbed gastrointestinally, the acute toxicity of orally administered aluminium compounds is relatively small (Table 7).

Compounds which dissolve well in water, are better absorbed than those which dissolve with difficulty in water. This is reflected in the LD_{50} values (dose which leads to the death of 50% of the laboratory animals treated), which were lower in the case of the well soluble aluminium chloride or nitrate than in the less soluble aluminium sulfate. No LD_{50} values are known for aluminium hydroxide, the most frequently used compound in the medical sector. However, it can be assumed that the hydroxide has a much lower toxicity because of its poor solubility in water than the soluble compounds described above. The toxic potency of systemically administered aluminium (Table 7) shows that this element, once it has passed the gastrointestinal barrier in larger quantities, can have serious biological effects.

In the case of subchronic oral administration, the main effect is a hypophosphatemia based on gastrointestinal phosphate binding to the aluminium cations: in rats

Table 7. LD_{50} values of several aluminium compounds [56]

Salt	Species	Route	LD_{50} mg Al^a/kg
Aluminium nitrate	Rat	oral	308
Aluminium chloride	Mice	oral	426
Aluminium sulfate	Mice	oral	979
Aluminium sulfate	Mice	ip^b	22
Aluminium nitrate	Mice	ip	23

^{a} equivalent to elemental aluminium (Al)
^{b} ip = intraperitoneal

which were fed with 6–10 mg Al/kg of body weight for 4 weeks, the growth decreased and rachitic bone changes occurred. In rats with phosphate additions to their diet, these symptoms were not seen. A reduction in the incorporation of radioactive phosphorus (^{32}P) in phospholipids, nucleic acids (RNA and DNA) was observed during oral aluminium administrations. Daily chronic administration of aluminium chloride corresponding to 36.5 mg Al/kg of body weight and day to rats for 55 days and single doses of 225 mg Al/kg of body weight caused changes in adenosine phosphates (increase of AMP and ADP and drop of ATP) in the serum [58].

Effects in the Cellular and Subcellular Sector

Aluminium seems to enter into interactions with cell membranes, the cytoskeleton and the genom, however, only in unphysiologically high concentrations: in an experiment by Bonhaus et al. [102] 0.1 mM (millimole) aluminium in vitro prevented the microtubuli polymerization of a protozoan. Vierstra and Haug [103] observed changes in the membrane lipid fluidity in another protozoan with 10^{-5}–10^{-2} M aluminium in the medium. De Boni et al. [104] reported that chromatin bound aluminium did not cause any ultrastructural changes in neurons of the spinal cord, in the case of cortex neurons, however, intracellular filaments accumulated and the survival of the neurons decreased. These changes occurred at an aluminium concentration in the medium of 3.3 µg/ml.

At a concentration of 2.3 µg/ml, the frequency of the "Sister Chromatide Exchanges" (SCE) in human lymphocytes in vitro declined. At a concentration of 26.4 µg/ml, the SCE frequency again reached the reference values. Furthermore an increased incorporation of tritiated thymidine in human astrocytes was observed at a concentration of 2.2 µg Al/ml. The DNA synthesis had been activated therefore [77]. In an experiment by Sanderson et al. [105] aluminium prevented the steroid induced chromosomes puffs of Simulium vittatum. The interactions of aluminium with DNA were examined in vitro at different pH values by Karlik and Eichhorn [98]. At pH 6, aluminium had a stabilizing effect on the DNA structure. This was measured on basis of the denaturing temperature of the double helix. At a lower pH, the presence of aluminium had a destabilizing effect. The authors then deduced the model of two aluminium DNA complexes: complex I, which is formed by the binding of Al(OH)$^{2+}$ to the phosphate groups with a stabilizing effect and complex II, at which Al^{3+} binds to the bases with a destabilizing effect.

Effects on the Blood Count

Trapp [96] proposes that the binding of aluminium at two iron binding sites of serum transferrine played a role for the development of microcytary anaemia in dialysis patients. Rats which were treated with aluminium salts [106], were affected by prophyria. Under the UV light, eyes, brain, bones and peritesticular fat fluoresce. Possible reasons for these observations could be the influence of aluminium on ferroxidase and δ-amino laevulinic acid dehydrase, two enzymes participating in blood formation. In the first case the function is adversely affected by a change in conformation, in the second case a reduction of the activity in vitro and a stimulation in vivo occurred [3, 107].

Effects on Bone Metabolism

Patients with chronic renal failure show high aluminium levels due to the treatment with aluminium containing phosphate binders or through contact with aluminium containing dialysate. They develop osteomalacy relatively frequently, which is a pathological demineralization of the bones. These patients have higher aluminium concentrations in the bone substance than dialysis patients without osteomalacy. Aluminium accumulates on the border between mineralized bones and osteoid [108]. Treatment with vitamin D_3 often proves to be unsuccessful in such cases. The rachitic changes in bones observed in tests on laboratory animals were mainly based on the hypophosphatemia caused by aluminium. Since in osteomalacic patients there is often no hypophosphatemia, direct toxic effects of aluminium on the bones were postulated.

Ellis et al. [69] in an experiment with animals avoided the gastrointestinal phosphate binding and therefore, a possible hypophosphatemia by administering aluminium intraperiotoneally: the doses were in the range of 38–109 mg aluminium spread over 45 to 85 days. After 53 days, the aluminium content in the bone ash had risen from an average 15.4 µg/g in reference animals to 109.3 µg/g in the animals treated. Nevertheless at that time there were no irregularities to be seen in the bones radiographically. After 63 days, however, osteomalacia developed. The aluminium content in the bone reached 163 µg/g at that time, which is in the concentration range of dialysis patients with osteomalacia. The concentrations of phosphorus and calcium in the blood and other blood parameters did not change in that period. The mechanisms of this direct toxic action are unknown: Lieberherr et al. [109] observed a stimulation of the acid and alkaline bone phosphatases by aluminium in vitro. This result is, however, contradictory to another in vitro examination by Sugawara et al. [110] in which neither a restriction nor a stimulation of alkaline and acid phosphatase by 10^{-8} to 10^{-4} M aluminium took place. Cadmium, which has been proved to cause osteomalacia (Itai-Itai disease), caused a marked decrease of the phosphatase in this experiment.

Neurotoxicity of Aluminium

The oldest studies based on experiments with animals in which a neurotoxic effect of aluminium was shown, stem from the 19th century [3]. Approximately 50 years later, aluminium compounds were first used to trigger off epilepsy experimentally. The discovery that aluminium causes epilepsy in animal experiments, was mere chance: originally the studies were conducted with proteins precipitated with aluminium which were applied to the surface of the brain.

It was ascertained later, however, that aluminium itself had a stronger effect than the proteins. Furthermore aluminium proved to be the strongest cause among 23 tested metals of chronic experimental epilepsy [111]. A few decades then passed until the neurotoxic potency of aluminium was investigated systematically. In such experiments with animals the administration of aluminium took place either intracerebrally as a single dose or subcutaneously in doses of 5 mg Al/kg of body weight daily for 19–30 days [77]. In rabbits, cats, monkeys and dogs, progressive encephalopathy was induced by aluminium. The symptoms were: changes in behaviour, muscular twitching, paroxisms, and death [77]. More recently Bugiani and Ghetti [112] also observed pathological changes in peripheral nerves and muscles as accompanying phenomena in encephalopathy. The typical morphological characteristic in the brain is a special type of neurofibrillary degeneration (NFD), which is detectable histologically as tangles in the cells [113]. The critical aluminium concentration, as from which an NFD appears, is 4 µg/g brain substance, at a normal concentration of 1–2 µg/g. In monkeys, with aluminium induced epilepsy, no NFD develops [77, 108]. Rats are even less sensitive: with aluminium concentrations in the brain which would be lethal for cats and rabbits, no signs of encephalopathy were found. Specific examinations revealed merely subtle changes in behaviour in rats [114] and effects of the following biochemical parameters: the ribosomal activity, the protein synthesis, the activity of the acetylcholine esterase and the intake of different neurotransmitters into the synaptosomes [77]. The neurotoxic activity of aluminium in sensitive laboratory animals could partially be based on interactions with phosphate, calcium, magnesium and enzymes: aluminium adversely affects the oxidative phosphorylization while reducing the ATP concentration. Furthermore the hexokinase activity in vitro with the formation of aluminium ATP in place of magnesium ATP was restricted; the high binding activity of aluminium to citrate, however, prevents this reaction taking place in vivo [115].

A reduction in the incorporation of ^{32}P in phospholipid, RNS and DNS was also observed [77].

In studies by Lai et al. [116] aluminium proved, in vitro, to be a relatively weak restrictor of the synaptosomal Na-K-ATP-ase and of the Mg-ATP-ase. 24 µM aluminium reduced the choline intake into the synapsomes by half. Wong and Lim [117] showed that the methylation of phospholipids into the synaptosomal membrane was lightly stimulated through aluminium. Ohtawa et al. [118] administered 100 mg $Al(OH)_3$/kg of body weight to rats for 7 days. This led to an increase in the lipid peroxide in the brain to 142% of the reference values. In other organs, the content of lipid peroxides remained unchanged. The activity of the superoxide dismutase was reduced in the brain and elevated in the kidneys. Siegel and Haug [99] demonstrated that the binding of aluminium to calmoduline caused structural changes and a blockage of the phosphodiesterase, which are dependent on calcium calmoduline.

Neuronal Disorders in Human Beings with Elevated Aluminium Content in the Brain

Elevated aluminium content in the brain was observed in the normal ageing process and in cases of various neuronal disorders in human beings, the causes of which are unknown. An etiological role of aluminium in the disorders was considered based on experiments with animals, which attribute aluminium with a neurotoxic function [77]. On the other hand aluminium could just represent a tracer of a disorder which shows up as symptom for a damaged blood brain barrier which gives aluminium greater access to the brain. It cannot be excluded, however, that the accumulated aluminium due to the primary degenerative processes in the brain accelerates the process of the disease secondarily [77]. Among the neuronal disorders with increased aluminium concentrations in the brain, one of the most important is the dementia of Alzheimer type (DAT) which occurs mainly in senility, but also presenility. The causes of DAT are unknown. The suggestion of genetic factors is supported by the recent detection of a gene on chromosome 21 which codes the synthesis of the brain amyloid proteins found in the brains of patients with Alzheimer's disease and Down's syndrome [119–122]. 30% of the brain of a DAT patient contains an aluminium concentration of more than 4 µg/g.

Such levels have been proved to be toxic in animal experiments. The brains of healthy reference people of the same age showed a content of 1–2 µg/g [77]. This result, however, is not uncontested. Various authors did not find any differences between DAT patients and reference individuals as regards the aluminium concentration. It was found later on that this contradiction was based on the sample size and case selection: since aluminium is not homogeneously distributed in the brain, but enriches locally, the critical areas cannot be recognized with too large sample quantities. Small sample quantitites ($\leq$ 100 mg wet weight) from different areas are therefore necessary [77]. The recent finding that the core of the characteristic plaques consists of amorphous alumino-silicate [123] has renewed the interest in the relationship of aluminium and Alzheimer's disease [124]. Aluminium has also been found in the neurofibrillary tangle-bearing neurons [125], another typical morphological feature in the brains of Alzheimer patients. Neurofibrillary tangles have also been observed in animals experimentally intoxicated with aluminium. The ultrastructure of these neurofilaments are single stranded, whereas they are paired and helically wound around each other in Alzheimer's disease [113]. They also show different staining properties [126]. While in the experiments with animals there was a high aluminium exposure, this is not the case in DAT. Blood plasma of DAT patients and other tissues than certain areas in the brain did not contain any higher quantities of aluminium. It has been put forward [127], however, that aluminium is transported to the cell surface bound to the iron transport protein transferrin where it enters the cell. Obviously no mechanism exists for releasing aluminium out of the cell. Since these processes seem to proceed extremely slowly, only longliving cells in the body, that is the neurons, can accumulate aluminium over the years at levels which could finally under certain conditions become toxic to the cells. These hypotheses have to be thoroughly tested before a causative role of aluminium in Alzheimer's disease can be deduced. Possibly, magnesium and calcium deficiencies together with a high intake of aluminium which occurred years ago in the people of Guam in the Western Pacific, created a situation for the development of Parkinsonism combined with dementia [128]. There are, however, other hypotheses for the etiology of the disease such as intake of the toxic aminoacid N-methylamino-L-alanine [129]. Likewise the epidemiologically found correlation between the prevalence of Alzheimer's disease and the aluminium content of drinking water [130, 131], which constitute only a very minor contribution to the daily aluminium intake, can hardly be accepted at the present state of knowledge as evidence for toxic properties of trace amounts of aluminium.

Another central nervous disorder with high aluminium concentrations in the brain occurs in dialysis patients — dialysis encephalopathy (DE). The difference from DAT is that in this case there is an elevated aluminium intake due to the use of aluminium containing phosphate binders and/or aluminium containing dialysis solutions and heavily impaired elimination of aluminium as a consequence of kidney damage. The disorder developed in Europe from 1976 to 1977 with an incidence of 600 in 100000 dialysis patients. The disease begins with speaking and writing difficulties, disturbances in the coordination of movement, myoclonic twitching and ends with characteristic changes in the EEG (electroencephalogramme), with attacks, dementia and death [77, 108]. Alfrey [132] showed that the aluminium concentration in the blood and various organs, including the brain, and there mainly in the grey substance, is higher than in dialysis patients without DE. Correlations between the frequency of the occurrence of DE and the dialysate aluminium content were observed [108, 132]. In various cases, due to a limited non administration of the phosphate binders, a rapid improvement in the DE disorder was attained [133]. Regression of the DE was also observed after kidney transplantations and desferrioxiamine treatment [132]. Based on this knowledge, aluminium is today generally considered to be the main cause of DE. The critical plasma aluminium concentration above which DE can occur, is between 100 and 200 µg/L [134]. However, little is known about the morphological changes and the distribution of aluminium in the brain of these patients. Important is the finding that the intranuclear aluminium concentrations are not higher, contrary to DAT or experimental encephalopathy, and that also no NFD occurs [77].

Occupational Exposure to Aluminium

Besides examination for fluoride exposure in the aluminium industry, studies on the effects of aluminium compounds and gases on exposed individuals are less frequent. In most recent studies blood and urine levels are determined in aluminium workers.

These results are summarized in the section "Absorption, Excretion and Retention". First medical examinations stem from the time of the second world war when production was raised. In these years specific lung disorders such as aluminium dust lung and the Shaver disorder or bauxite lung are described [135]. The lung disorder which arose in bauxite plants, was a progressive, interstitial lung fibrosis of the non-nodular type, accompanied by marked emphysemes. Although exposure to aluminium oxide was high at such workplaces, the etiology was often questioned since, besides exposure to aluminium oxide, a considerable exposure to silicon oxide also existed. Today "aluminosis" is an occupational disorder which is entitled to payment of damages in some countries, although today cases of lung fibrosis in the production and processing of aluminium are hardly ever found due to improvement in worker's hygiene or to an inherent inability of aluminium oxide to produce lung fibrosis. The results of experiments with animals are inconclusive: fibroses were found occasionally in animals with high doses of aluminium powder [135], implantations of thin aluminium wires [136] and with fine particulate aluminium oxyhydrate [135]. With aluminium fibres [137] and aluminium chlorhydrate [79] only macrophage reactions were attained. Accordingly it has to be taken into account that in the different working processes in aluminium production and fabrication, different dust and gas emissions with possibly markedly diverging biological acitivities arise. This is also expressed in the different limit values ([138–140] Table 8).

In 1985 a value of 170 µg Al/L urine was fixed as a biological work tolerance value in Germany [141]. This value is based on a study of Mussi et al. [89] due to a correlation of aluminium elimination and aluminium concentrations in the air and the American threshold [139].

The European Community published guidelines for three levels of aluminium serum [142]: > 60 µg/L are indicative for increased aluminium absorption; > 100µg/L are indicative for a potential clinical concern; > 200 µg/L should never be exceeded because that would generally lead to clinical symptoms.

As regards the health effects of the aluminium oxide, one has to distinguish between the different aluminium oxides, which differ in their crystal structures. They are

Table 8. Limit values referring to airborne concentrations of aluminium and aluminium compounds

TLV-TWA (ACGIH)	1987–1988	[139]
α-Alumina	10 mg/m³	total dust (general threshold limit for nuisance particulates containing < 1 % quartz)
Aluminium		
Metal oxide	10 mg/m³	
Pyro powders	5 mg/m³	
Welding fumes	5 mg/m³	
Soluble salts	2 mg/m³	
Alkyls (NOC)[1]	2 mg/m³	[1] NOC = Not otherwise classified
DFG Maximum concentrations at the workplace (1988) [140]		
Aluminium		
Metal, oxide, hydroxide	6 mg/m³ [2,3]	
Aluminium oxide fume	6 mg/m³ [2,3]	
Short-term level	5 times MAC (= 30 mg/m³) during 30 min. with a frequency of 2 per shift	

[2] General Treshold Limit Value for fine dust with quartz content < 1 %.
[3] Long-term value for exposure lasting 1 year.

designated all together as Al_2O_3. Hydrated aluminas change to α Al_2O_3 with a hexagonal (rhombohedral structure) at about a calcination temperature of 1000–1200 °C from different types of initial compounds (gibbsite, boehmite, bayerite). When heating up to about 800 °C, different low temperature transitional aluminas which are partially poorly crystallized, are formed; the η, χ and γ-crystal structures. In the industrial nomenclature there is often no differentiation between these three forms. From about 800–1100 °C the $\varkappa$, δ and θ modifications are present [143]. Due to this variety and the practical difficulty of determining the modifications exactly, specially the non γ-modifications, there is some confusion about possible health effects due to the individual crystal modification.

While α-aluminium oxide is considered to be relatively inert, the so-called γ-aluminium oxide is attributed a certain fibrinogeneous potency [135]. It must be kept in mind, however, that in that study a laboratory grade material with very small particle size of 0.004–0.005 μm was used. This material differed in many respects from that used in the aluminium industry for reduction.

Ess et al. [144] have shown in a new study that for a short time after application of different smelter grade aluminas into the lung of animals, reactions to foreign bodies do occur but that these reactions did not lead to a fibrosis whereas the fibrogenic properties of laboratory grade materials could be confirmed.

During recent years reports on possible health effects of aluminium in occupational situations appeared only sporadically.

In an epidemiological study by Chan-Yeung et al. [145], spirometric and lung radiological tests on potroom workers were made. Highly exposed workers showed a higher susceptibility to coughing and a lower vital capacity of the lung. Asthma did not occur. The reasons for these effects are, besides irritations from contaminated aluminium compounds, also gases such as HF and SO_2.

Slightly exposed workers differ in no parameter from the reference groups. Fluoride content and the total dust quantity in the air were lower than the acceptable TLV. In another epidemiological study Chan-Yeung et al. [146] did not observe any effect on the liver and kidney function in any of the workers in aluminium casthouses.

Clonfero et al. [147] concluded in a paper based on examinations of 197 potroom workers, that the workers are exposed to an increased risk of bronchitis and pneumoconiosis when employed over a number of years and suggested the designation "aluminosis minima". The same group found slight radiological lung changes among potroom workers exposed to a total dust load of 3.4–6.5 mg/m^3 [148]. Due to possible prior exposure of the workers to quartz-containing dusts, the role of aluminium is still to be determined. In a welder from the aircraft building industry, a granulomatosis was diagnosed. Microanalytically Chen et al. [149] found aluminium to be the sole metal in large quantities of lung tissue.

Another case of a lung fibrosis in an aluminium arc welder was attributed by Vallyathan et al. [150] to the aluminium exposure. However, it was doubted by Cole et al. [151], the reason being that in the aluminium arc welder other possible fibrogeneous substances such as for example silicon compounds and ozone occur. There are some case reports of lung fibrosis possibly due to aluminium oxide exposure e.g. in the production of aluminium sulfate from hydrated aluminium oxide [152] and the inhalation of fibres of aluminium silicate mullite [153]. Such cases only occurred occasionally. The causal relationship of aluminium could not be shown unequivocally.

As regards central nervous disturbances, only very few reports are known e.g. [154], which were connected with exposure to aluminium, other causes than aluminium were not excluded in any of these reports [83]. At any rate, disorders of the central nervous system seem to be even less significant than the doubtful lung effects.

Acknowledgement: The authors thank Mrs. G. Stiffler, Mr. R. Pawlek and Mrs. D. Schär for their assistance.

References

1. Campbell IR, Cass JS, Cholak J, Kehce RA (1957) AMA Arch Ind Health 15: 359
2. Grjotheim K, Krohn C, Malinovsky M, Matiasovsky K, Thonstad J (1982) Fundamentals of Hall-Héroult Process. In: Aluminium Electrolysis 2 ed., Aluminium-Verlag, Düsseldorf
3. Sorenson JRJ, Campbell IR, Tepper LB, Lingg RD (1974) Environ Health Persp 8: 3
4. Anonymous (1989) Metal Bulletin No. 7354: 7
5. Pawlek R (1986, 1989) Primary Aluminium Smelters and Producers of the World, Aluminium-Verlag, Düsseldorf
6. Anonymous (1988) Metal Bulletin No. 7335: 13
7. Pawlek R (1987, 1989) Alumina Refineries and Producers of the World, Aluminium-Verlag, Düsseldorf
8. Dinman BD (1983) Aluminium, alloys, compounds. In: Encyclopaedia of occupational health and safety 3rd revised ed., International Labour Office, Geneva, p 131
9. Hem JD, Roberson CE (1967) Form and stability of aluminium hydroxide complexes in dilute solution, US Geol Surv Water-Supply Paper 1827-A
10. Bohn HL, McNeal BL, O'Connor GA (1979) Soil chemistry, John Wiley & Sons, New York, p 199
11. Burrows WD (1977) CRC Critical Rev Environ Control 7: 167
12. Marion GM, Hendricks DM, Dutt GR, Fuller WH (1976) Soil Sci 121: 76

13. Norton SA (1982) Proc int symp acidic precipitation and fishery impacts in Northeastern North America, Am Fis Soc, Bethesda, p 93
14. Lind CJ, Hem JD (1975), Effects of organic solutes on chemical reactions of aluminium, US Geol Surv Water-Supply Paper 1827-G
15. Chen YSR, Butler JN, Stumm W (1973) J Coll Interf Sci 43: 421
16. May HM, Helmke PA, Jackson ML (1979) Chem Geol 24: 259
17. Hydes DJ, Liss PS (1977) Estuar Coast Mar Sci 5: 755
18. Borg H (1983) Hydrobiol 101: 27
19. Budd WW, Johnson AH, Huss JB, Turner RS (1981) Wat Resour Res 17: 1179
20. Johnson NM, Driscoll ChT, Eaton JS, Likens GE, McDowell WH (1981) Geochim Cosmochim Acta 45: 1421
21. Dickson W (1980) Proc int conf ecol impact acid precip, SNSF project, Norway, p 75
22. Evans LS, Hendrey GR, Stensland GJ, Johnson DW, Francis AJ (1981) Wat Air Soil Poll 16: 469
23. Cronan CS, Schofield CL (1979) Science 204: 304
24. Driscoll ChT, Baker JP, Bisogni JJ, Schofield CL (1980) Nature 284: 161
25. Bjärnborg B (1983) Hydrobiol 101: 19
26. Bowen HJM (1979) Environmental chemistry of the elements, Academic Press, London, p 238
27. Webber MD, Hoyt PB, Nyborg M, Corneau D (1977) Can J Soil Sci 57: 361
28. MacLean AJ, Halstead RL, Finn BJ (1972) Can J Soil Sci 52: 427
29. Wada SI, Wada K (1980) J Soil Sci 31: 457
30. Pionke HB, Corey RB (1967) Soil Sci Soc Amer Proc 31: 749
31. Ulrich B, Mayer R, Khanna PK (1980) Soil Sci 130: 193
32. Bhumbla DR, McLean EO (1965) Soil Sci Soc Proc, p 370
33. Gonzalez-Erico E, Kamprath EJ, Naderman GC, Soares WV (1979) Soil Sci Soc Amer J 43: 1155
34. Johnson DW, Turner J, Kelly JM (1982) Wat Resour Res 18: 449
35. Foy CD, Chaney RL, White MC (1978) Ann Rev Plant Physiol 29: 511
36. MacLeod LB, Jackson LP (1967) Can J Soil Sci 47: 203
37. MacLeod LB, Jackson LP (1965) Can J Soil Sci 45: 221
38. Tischner R, Kaiser U, Hüttermann A (1983) Forstwissenschaftl Zentralblatt 102: 329
39. Muniz IP, Leivestad H (1980) Proc int conf ecol impact acid precip, SNSF project, Norway, p 84
40. Eriksson F, Hörnström E, Mossberg P, Nyberg P (1983) Hydrobiol 101: 145
41. Baker JP, Schofield CL (1982) Water, Air, Soil Pol 18: 289
42. Dietrich DR (1988) Aluminium toxicity to salmonids at low pH, Thesis No. 8715 Swiss Fed Inst Technol, Zurich
43. Schofield CL, Trojnar JR (1980) In: Toribara TY, Miller MW, Morrow PE (eds) Polluted rain, Plenum, New York and London p 341
44. Hunter JB, Ross SL, Tannahill J (1980) Wat Pollut Control 79: 413
45. Dietrich DR, Schlatter Ch (1989) Aquatic Toxicol (in press)
46. Brown DJA (1983) Bull Environ Contam Toxicol 30: 582
47. Baker JP (1982) Proc int symp acidic precipitation and fishery impact in Northeastern North America, Am Fish Soc, Bethesda, p 165
48. Petrich SM, Reish DJ (1979) Bull Environ Contam Toxicol 23: 698
49. Biesinger KE, Christensen GM (1972) J Fish Res Board Can 29: 1691
50. Schafran GC, White JR, Driscoll ChT (1982) Northeast Environ Sci 1: 151
51. Dietrich D, Schlatter Ch, Blau N, Fischer M (1989) Toxicol Environm Chem 19: 17
52. Olsson H (1983) Hydrobiol 101: 49
53. Treier S, Kluthe R (1988) Ernährungs-Umschau 35: 307
54. Jackson ML, Ming Huang P (1983) Sci Total Environ 28: 269
55. Attanandana T, Vacharotayan S, Kyuma K (1982) In: Dost H, van Breemen N (eds) Proc 2nd int symp on acid sulphate soils, Int inst land reclam improv, Wageningen, NL, p 137
56. Rickenbacher U (1984) Mitt Gebiete Lebensm Hyg 75: 16
57. Lione A (1983) Fd Chem Toxic 21: 103
58. WHO (1977) Food Additives Series No. 12 and WHO (1982) Technical Report Series No. 683
59. Knutti R, Zimmerli B (1985) Mitt Gebiete Lebensm Hyg 76: 206
60. Flaten TP, Ødegård M (1988) Fd Chem Toxic 26: 959
61. Coriat AM, Gillard RD (1986) Nature 321: 570
62. Greger JL, Baier MJ (1983) Fd Chem Toxic 21: 473

63. Mayor GH, Keiser JA, Ku PK (1977) Science 197: 1187
64. Feinroth M, Feinroth MV, Berlyne GM (1982) Miner Electrol Metab 8: 29
65. Kaehny WD, Hegg AP, Alfrey AC (1977) N Engl J Med 296: 1389
66. Galle P (1981) CR Séances Acad Sci Ser 3, p 91
67. Gorsky JE, Dietz AA, Spencer H, Osos D (1979) Clin Chem 25: 1739
68. Marquis JK (1982) Bull Environ Contam Toxicol 29: 43
69. Ellis HA, McCarthy JH, Herrington J (1979) J Clin Pathol 32: 832
70. Alfrey AC, Hegg A, Craswell P (1980) Amer J Clin Nutr 33: 1509
71. Cann CE, Prussin SG, Gordan GS (1979) J Clin Endocrin Metabol 49: 543
72. Harwerth A, Kufner G, Helbing F (1987) Arbeitsmed Sozialmed Präventivmed 22: 2
73. Hartung M, Schaller KH, Kentner M, Manke HG, Weltle D, Grimm HG (1984) In: Bericht über 24. Jahrestagung der Deutschen Gesellschaft für Arbeitsmedizin e.V., Mainz, p 289
74. Yukawa M, Suzuki-Yasumoto M, Amano K, Terai M (1980) Arch Environ Health 35: 36
75. Flendrig JA, Kruis H, Das HA (1976) Lancet I (7971): 1235
76. Chan YL, Alfrey AC, Posen S, Lissner D, Hills E, Dunstan CR, Evans RA (1983) Calcif Tissue Int 35: 344
77. Crapper DR, De Boni U (1980) In: Liss L (ed) Aluminium neurotoxicity, Pathotox, Park Forest South, p 3
78. Arieff AI, Cooper JD, Armstrong D, Lazarowitz VC (1979) Ann Int Med 90: 741
79. Stone CJ, McLaurin DA, Steinhagen WH, Cavender FL, Haseman JK (1979) Toxic Appl Pharma 49: 71
80. Capel ID, Pinnock MH, Dorell HM, Williams DC, Grant CG (1981) Clin Chem 27: 879
81. Hall GS, Carr MJ, Cummings E, Lee M (1983) Clin Chem 29: 1318
82. Shore D, Wyatt RJ (1983) J Nerv Ment Dis 171: 553
83. McLaughlin AIG, Kazantzis G, King E, Teare D, Porter RJ, Owen R (1962) Brit J Industr Med 19: 253
84. Sjögren B, Lundberg I, Lidums V (1983) Br J Ind Med 40: 301
85. Sjögren B, Lidums V, Håkansson M, Hedström L (1985) Scand J Work Environ Health 11: 39
86. Schlatter Ch, Steinegger A, Rickenbacher U, Hans Ch, Lengyel A (1986) Soz Präv Med 31: 125
87. Valentin H, Preusser P, Schaller KH (1976) Int Arch Occup Environ Health 38: 1
88. Schaller KH, Valentin H (1984) In: Alessio L, Berlin A, Boni M, Roi R (eds) Biological indicators for the assessment of human exposure to industrial chemicals, Commission of the European Communities, p 21
89. Mussi I, Calzaferri G, Buratti M, Alessio L (1984) Int Arch Occup Environ Health 54: 155
90. Leung FY, Henderson AR (1983) Atomic Spectr 4: 1
91. Alderman FR, Gitelman HJ (1980) Clin Chem 26: 258
92. Ward MK, Feest TG, Ellis HA, Parkinson IS, Kerr DNI, Herrington J, Goode GLT (1978) Lancet I 8069: 841
93. Pegon Y (1978) Anal Chim Acta 101: 385
94. Schramel P, Wolf A, Klose BJ (1980) J Clin Chem Biochem 18: 591
95. Elliot; cit. in King SW, Savory J, Wills MR (1981) Crit Rev Lab Sci 14: 16
96. Trapp GA (1983) Life Sci 33: 311
97. Galle P, Chatel M, Berry JP, Menault F (1979) Nouv Presse Med 8: 4091
98. Karlik SJ, Eichhorn GL (1980) In: Liss L (ed) Aluminium neurotoxicity, Pathotox, Park Forest South, p 93
99. Siegel N, Haug A (1983) Biochim Biophys Acta 744: 36
100. Mazarguil H, Haran R, Laussac JP (1982) Biochim Biophys Acta 717: 465
101. Maloney NA, Ott SM, Alfrey AC, Miller NL, Coburn JW, Sherrard DJ (1982) J Lab Clin Med 99: 206
102. Bonhaus DW, McCormack KM, Mayor GH, Mattson JC, Hook JB (1980) Toxicol Lett 6: 141
103. Vierstra R, Haug A (1978) Biochem Biophys Res Comm 84: 138
104. DeBoni U, Seger M, Crapper DR (1980) In: Liss L (ed) Aluminium neurotoxicity, Pathotox, Park Forest South, p 65
105. Sanderson CL, McLachlan DRC, De Boni U (1982) Can J Gen Cytol 24: 27
106. Berlyne GM, Yagil R (1973) Lancet 29: 1501

107. Abdulla M, Svensson S, Haeger-Aronsen B (1979) Archives Environ Health 34: 464
108. King SW, Savory J, Wills MR (1981) Crit Rev Clin Lab Sci 14: 1
109. Lieberherr M, Grosse B, Cournot-Witmer G, Thil CL, Balsan S (1982) Calcif Tissue Int 34: 280
110. Sugawara N, Sadamoto T, Sugawara C (1983) Bull Environ Contam Toxicol 31: 386
111. Kopeloff LM, Barrera SE, Kopeloff N (1942) Amer J Psych 98: 881
112. Bugiani O, Ghetti B (1982) Neurobiol Aging 3: 209
113. Wisniewski HM, Soifer D (1979) Mechanisms of Aging and Development 9: 119
114. King GA, De Boni U, Crapper DR (1975) Pharmacol Biochem Behav 3: 1003
115. Martin RB (1986) Clin Chem 32: 1797
116. Lai JCK, Guest JF, Leung TKC, Lim L, Davison AN (1980) Biochem Pharm 29: 141
117. Wong PCL, Lim L (1981) Biochem Pharm 30: 1704
118. Ohtawa M, Seko M, Takayama F (1983) Chem Pharm Bull 31: 1415
119. Kang J, Lemaire HG, Unterbeck A, Salbaum JM, Masters CL, Grzeschik KH, Multhaup G, Beyreuther K, Müller-Hill B (1987) Nature 325: 733
120. Goldgaber D, Lerman MI, McBride OW, Saffiotti U, Gajdusek DC (1987) Science 235: 877
121. Tanzi RE, Gusella JF, Watkins PC, Bruns GAP, St George-Hyslop P, Van Keuren ML, Patterson D, Pagan S, Kurnit DM, Neve RL (1987) Science 235: 880
122. St George-Hyslop PH, Tanzi RE, Polinsky RJ, Haines JL, Nee L, Watkins PC, Myers RH, Feldman RG, Pollen D, Drachman D, Growdon J, Bruni A, Foncin JF, Salmon D, Frommelt P, Amaducci L, Sorbi S, Piacentini S, Stewart GD, Hobbs WJ, Conneally PM, Gusella JF (1987) Science 235: 885
123. Candy JM, Oakley AE, Klinowski J, Carpenter TA, Perry RH, Atack JR, Perry EK, Blessed G, Fairbairn A, Edwardson JA (1986) Lancet I (8477): 354
124. Candy JM, Edwardson JA, Klinowski J, Oakley AE, Perry EK, Perry RH (1985) In: Traber J, Gispen WH (eds) Senile dementia of the Alzheimer type, Springer, Berlin-Heidelberg
125. Perl DP, Brody AR (1980) Science 208: 297
126. Wisniewski HM, Wen GY (1985) Acta Neuropathol 66: 173
127. Birchall JD, Chappell JS (1988) Lancet II (8618): 1008
128. Garruto RM, Yase Y (1986) Trends in Neurosciences 9: 368
129. Kurland LT (1988) Trends in Neurosciences 11: 51
130. Martyn CN, Barker DJP, Osmond C, Harris EC, Edwardson JA, Lacey RF (1989) Lancet I (8629): 59
131. Flaten TP (1986) An investigation of the chemical composition of Norwegian drinking water and its possible relationships with the epidemiology of some diseases. Thesis, Teknikse Høgskole Trondheim
132. Alfrey AC (1982) Proc int workshop on the role of biological monitoring in the prevention of aluminium toxicity in man, Commission of the European Communities, Luxembourg, p 1
133. McKinney TD, Basinger M, Dawson E, Jones MM (1982) Nephron 32: 53
134. Minutes and summary report (1982) Int workshop on the role of biological monitoring in the prevention of aluminium toxicity in man, Commission of the European Communities, Luxembourg
135. Klosterkötter W (1960) AMA Arch Ind Health 21: 458
136. Greenberg SR (1977) Lab Invest 36: 339
137. Gross P, de Treville RTP, Cralley LJ, Grandquist WT, Pundsack FL (1970) Amer Ind Hyg Assoc J 31: 125
138. Documentation of the threshold limit values, 4th edn, American conf of governmental industrial hygienists Inc, Cincinnati, Ohio
139. Threshold limit values and biological exposure indices for 1987—1988 (1987) American conf of governmental industrial hygienists Inc, Cincinnati, Ohio
140. DFG, Deutsche Forschungsgemeinschaft (1988) Maximale Arbeitsplatzkonzentrationen und Biologische Arbeitsstofftoleranzwerte, Verlag Chemie, Weinheim
141. DFG, Deutsche Forschungsgemeinschaft (1986) BAT-Werte, Arbeitsmedizinisch-toxikologische Begründungen. Bd 1, Aluminium, Verlag Chemie, Weinheim, in preparation
142. Anonymous (1982) J Clin Chem Clin Biochem 20: 837
143. Misra S (1986) Industrial alumina chemicals, ACS Monograph 184, Am Chem Soc, Washington DC
144. Ess S, Steinegger A, Ess HJ, Rüttner JR, Schlatter Ch in preparation

145. Chan-Yeung M, Wong R, MacLean L, Tan F, Schulzer M, Enarson D, Martin A, Dennis R, Grzybowski S (1983) Am Rev Resp Dis 127: 465
146. Chan-Yeung M, Wong R, Tan F, Enarson D, Schulzer M, Ostrow D, Knickerbocker J, Subbarao K, Grzybowski S (1983) Arch Environ Health 38: 34
147. Clonfero E, Cortese S, Saia B, Marcer G, Crepet M (1978) Med Lavoro 69: 613
148. Saia B, Cortese S, Piazza G, Camposampietro A, Clonfero E (1981) Med Lavoro 72: 323
149. Chen WJ, Monnat RJ, Chen M, Mottet NK (1978) Hum Path 9: 705
150. Vallyathan V, Bergeron WN, Robichaux PA, Craighead JE (1982) Chest 81: 372
151. Cole HM, Benton RE, Skalsky HL (1983) Chest 83: 291
152. Cataldi R, De Palma M, Valente T, Bonsignore AD (1981) Med Lavoro 72: 318
153. Golden E, Warnock M (1981) Am Rev Resp Dis 123: 129
154. Longstreth WT, Rosenstock L, Heyer NJ (1985) Arch Intern Med 145: 1972

Subject Index

The Handbook of
**Environmental
Chemistry**
Edited by O. Hutzinger

Volume 2

Reactions and Processes

Part D

1988. XI, 210 pp. 47 figs. 55 tabs.
ISBN 3-540-15547-3

Contents: *R. Herrmann,* Bayreuth, FRG: Hydrology. –
N. O. Crossland, Sittingbourne, UK; *C. J. M. Wolff,* Amsterdam,
The Netherlands: Outdoor Ponds: Their Construction,
Management, and Use in Experimental Ecotoxicology. –
T. Mill, Menlo Park; *W. Mabey,* San Francisco, USA: Hydrolysis
of Organic Chemicals. – *M. Waldichuk,* Vancouver, Canada:
Exchange of Pollutants and Other Substances Between the
Atmosphere and the Oceans. – *P. B. Tinker,* Swindon;
P. B. Barraclough, Harpenden, UK: Root-Soil Interactions. –
C. M. Menzie, Washington, D.C., USA: Reaction Types in the
Environment.

Part E

1989. 256 pp. 50 figs. 40 tabs.
ISBN 3-540-51126-1

Contents: *R. P. Wayne,* Oxford, UK: The Photochemistry of
Ozone. – *B. L. Worobey,* Ottawa, Canada: Nonenzymatic Bio-
mimetic Oxidation Systems: Theory and Application to Trans-
formation Studies of Environmental Chemicals. –
J. L. M. Hermens, Utrecht, The Netherlands: Quantitative Struc-
ture-Activity Relationships of Environmental Pollutants. –
A. Opperhuizen, D. T. H. M. Sijm, Amsterdam, The Netherlands:
Biotransformation of Organic Chemicals by Fish: Enzyme
Activities and Reactions.

Volume 4

Air Pollution

Part B

1989. XI, 262 pp. 93 figs. ISBN 3-540-50915-1

Contents: *H. Brauer,* Berlin: Air Pollution Control Equipment.
– *J. S. Gaffney, N. A. Marley,* Los Alamos, NM; *E. W. Prestbo,*
Seattle, WA, USA: Peroxyacyl Nitrates (Pans): Their Physical
and Chemical Properties. – *R. Harkov,* Somerset, NJ, USA:
Semivolatile Organic Compounds in the Atmosphere. –
F. W. Lipfert, Northport, NY, USA: Air Pollution and Materials
Damage. – *G. E. Shaw,* Fairbanks, AK; *M. A. K. Khalil,*
Beaverton, OR, USA: Arctic Haze.

Springer-Verlag Berlin
Heidelberg New York London
Paris Tokyo Hong Kong

Springer

Environmental Toxin Series

Editors-in-Chief: O. Hutzinger,
University of Bayreuth, FRG;
S. H. Safe, Texas A&M University,
College Station, TX, USA

Editorial Board: M. W. Anders,
J. DiGiovanni, P. S. Guzelian,
A. W. Hayes, M. A. Hayes, G. W. Ivie,
R. Koch, H. J. Lewerenz, E. Löser,
J. D. McKinney, A. Parkinson,
T. D. Phillips, I. G. Sipes, J. Thies,
H. R. Witschi

The concern about environmental
toxins is ever increasing, as is the need
for sound scientific information. The
Environmental Toxin Series is dedicated
to the publication of comprehensive
reviews and monographs on compounds
or classes of chemicals which are of
importance in environmental toxicology. The series is designed to serve as a
background of information for scientific
investigation as well as risk analysis and
political decision making. The main aim
of the series is to describe in as
complete a way as possible all potentially hazardous chemicals from the
point of view of chemistry, ecology,
toxicology, risk analysis and regulatory
implications. From time to time confer-
ence proceedings on inportant and
urgent topics will be included in the
series. We thank the members of the
editorial board for their enthusiastic
support.

Springer-Verlag Berlin
Heidelberg New York London
Paris Tokyo Hong Kong

Volume 1

S. Safe, Texas A&M University, College
Station, TX, USA (Ed.)

Polychlorinated Biphenyls (PCBs): Mammalian and Environmental Toxicology

1987. X, 152 pp. 33 figs. 35 tabs.
ISBN 3-540-15550-3

Contents: *S. Safe, L. Safe, M. Mullin:*
Polychlorinated Biphenyls: Environmental
Occurrence and Analysis. – *L. G. Hansen:*
Environmental Toxicology of Polychlori-
nated Biphenyls. – *A. Parkinson, S. Safe:*
Mammalian Biologic and Toxic Effects of
PCBs. – *M. A. Hayes:* Carcinogenic and
Mutagenic Effects of PCBs. – *I. G. Sipes,*
R. G. Schnellmann: Biotransformation of
PCBs: Metabolic Pathways and Mecha-
nisms. – *R. J. Lutz, R. L. Dedrick:* Physio-
logic Pharmacokinetic Modeling of
Polychlorinated Biphenyls. – *S. Safe:* PCBs
and Human Health. – Subject Index.

Volume 2

M. Stoeppler, Jülich, FRG; **M. Piscator,**
Stockholm, Sweden (Eds.)

Cadmium
3rd IUPAC Cadmium Workshop, Juelich, FRG, August 1985

1988. XII, 237 pp. 57 figs. 79 tabs.
ISBN 3-540-15551-1

Contents: I. Toxicity, Carcinogenicity,
Animal Experiments – II. Epidemiology –
III. Cadmium in the Environment – IV.
Methodology and Quality Assessment.

Springer